# UNDERSTANDING THE MANUFACTURING PROCESS

# MANUFACTURING ENGINEERING
# AND MATERIALS PROCESSING

*A Series of Reference Books and Textbooks*

SERIES EDITORS

### Geoffrey Boothroyd

*Department of Mechanical Engineering*
*University of Massachusetts*
*Amherst, Massachusetts*

### George E. Dieter

*Dean, College of Engineering*
*University of Maryland*
*College Park, Maryland*

1. Computers in Manufacturing, *U. Rembold, M. Seth, and J. S. Weinstein*
2. Cold Rolling of Steel, *William L. Roberts*
3. Strengthening of Ceramics: Treatments, Tests, and Design Applications, *Henry P. Kirchner*
4. Metal Forming: The Application of Limit Analysis, *Betzalel Avitzur*
5. Improving Productivity by Classification, Coding, and Data Base Standardization: The Key to Maximizing CAD/CAM and Group Technology, *William F. Hyde*
6. Automatic Assembly, *Geoffrey Boothroyd, Corrado Poli, and Laurence E. Murch*
7. Manufacturing Engineering Processes, *Leo Alting*
8. Modern Ceramic Engineering: Properties, Processing, and Use in Design, *David W. Richerson*
9. Interface Technology for Computer-Controlled Manufacturing Processes, *Ulrich Rembold, Karl Armbruster, and Wolfgang Ülzmann*
10. Hot Rolling of Steel, *William L. Roberts*
11. Adhesives in Manufacturing, *edited by Gerald L. Schneberger*
12. Understanding the Manufacturing Process: Key to Successful CAD/CAM Implementation, *Joseph Harrington, Jr.*

*OTHER VOLUMES IN PREPARATION*

# UNDERSTANDING THE MANUFACTURING PROCESS

## Key to Successful CAD/CAM Implementation

**JOSEPH HARRINGTON, JR.**
Consulting Engineer

Consultant to:
*Arthur D. Little, Inc.*
*Cambridge, Massachusetts*

MARCEL DEKKER, INC.        New York and Basel

Library of Congress Cataloging in Publication Data

Harrington, Joseph, [date]
   Understanding the manufacturing process.

   (Manufacturing engineering and materials processing;
12)
   Includes index.
   1. Production management.  2.  CAD/CAM systems.
I. Title.  II. Series.
TS155.H2953    1984      658.5      84-4323
ISBN 0-8247-7170-2

MARCEL DEKKER, INC.
270 Madison Avenue, New York, New York  10016

Current printing (last digit):
10  9  8  7  6  5  4  3  2  1

PRINTED IN THE UNITED STATES OF AMERICA

# Foreword

This book goes to press at a time when manufacturing in the USA is under the duress of new industrial competition, especially from Japan and other Asian nations, but indeed coming from vigorous competitors the world over. American manufacturers have lost market shares not only for reasons of cost but, clearly, also because of poor quality, slow new product development, lack of reliability, and limited service.

Between 1960 and 1980 the most vaunted and powerful industrial system the world has ever seen was challenged for the first time. Events have proved in many industries that our industrial system and its management were not as good as we had believed. Before 1960 there was little serious and organized competition. After 1960, when the nature of competition changed, we caved in with surprisingly little resistance. Our quick slide has raised some questions: Was our industrial management really as good as we thought all along? Are the knowledge and wisdom we have built up and accumulated over 200 years "valid" – or are they getting us into trouble?

There is considerable evidence that our wisdom about manufacturing management maybe is not all that wise. The Japanese have challenged our assumptions and premises about inventory, about the role of the first-line supervisor, about the proper roles of workers, about gear-ing up for the introduction of new products, about vendor relationships, about set-up and changeover times, and about maintenance. Hundreds of experienced industrial managers visit Japan, Taiwan, Germany, and Switzerland to return confounded and surprised that what they had learned as truth here was not true everywhere. Our recent history of industrial malaise has indeed a long-term more-or-less continual antecedent of employee malaise with strikes and withheld effort and commitment dating back to our first textile mills. It all now makes us start to wonder about how to manage manufacturing – all over again.

For years, seemingly forever, we have thought of managing manufacturing as controlling the effort and actual physical processes of making things. Frederick Taylor taught managers to analyze in depth and with great care exactly what each worker did on or to the product and he urged management to employ some clerks and staff to get the work ready and count it and plan it and control it. The focus of industrial management has not only been on the physical side of manufacturing but equally on the cost of doing the physical operations.

Now, in this time of self-reexamination, many are beginning to wonder whether, all along, our focus and emphasis have not been misdirected. Industrial management may have been too much directed at physical productivity and in that preoccupation most companies have slowly but steadily added indirect employees and costs and overhead and paperwork.

Joseph Harrington comes forth in this book with a refreshing new look at this old art of management. He depicts manufacturing in terms of the basic steps which inherently always need to be done and he describes them clearly. However he places the whole panoply of manufacturing activities in a most unconventional new conceptual plane – that of information, rather than the usual focus on physical transformations and costs.

As readers move through the book, they may ask—now, if we look at manufacturing in the way Dr. Harrington suggests, and redesign our manufacturing systems to focus on accurate and timely information, would it not perhaps help us to meet the new industrial competition – which is essentially technologically based – by becoming organizations which learn faster and adapt better?

Wickham Skinner  
James E. Robison Professor  
Harvard Business School  
Boston, Massachusetts

# Foreword

Dr. Joseph Harrington has written the definitive book critical to the problem undermining American manufacturing industry today – the failure to understand the real nature of manufacturing in *today's* world. The strength of the American manufacturing industry used to be its excellent understanding of manufacturing and how to make it work successfully – when it depended wholly on "people-to-people data exchange modes." The problem is, as the author illustrates so well in his Preface, that the industry is still trying to operate in that same old bits-and-pieces manner, superimposing bits and pieces of manufacturing-related computer technology on top of it. He shows them that they must understand and deal with the whole cloth of manufacturing – the continuum of manufacturing and all its interrelationships – which can be so expertly woven together by that computer technology today.

Not only is Dr. Harrington's book critical to the American manufacturing industry per se, it is also critical to American engineering education. In recent years, American universities have practically forgotten that manufacturing even exists. Now with the industry crying for help, they are scrambling to try to understand and teach what modern manufacturing is all about. This book supplies just what they need, both to understand and to teach.

The author's use of a model to do this – the ICAM IDEF model – is certainly *the* way to help people understand. One picture is not only worth a thousand words; a thousand words are worth *much* more when reinforced with a picture. It is sad indeed that, again, so little of the American manufacturing industry has appreciated the power and value of the IDEF methodology for understanding and designing their manufacturing enterprises and operations for effective use of computer technology – even though that methodology was originated and developed right in this country. Now, at last, this book shows them in a straightforward step-by-step hand-holding way just how it applies and how to use it. That is indeed a tremendous service!

M. Eugene Merchant
Metcut Research Associates, Inc.
Cincinnati, Ohio

# Preface

Some time ago my grandson, Robert, was working on a jigsaw puzzle. One of the pieces fell accidentally to the floor, and the dog ate it. It was bad taste on the part of the dog, but the task of making a replacement part fell to me. I cut a piece of plywood of the right thickness to the exact convoluted outline needed to match the surrounding pieces. While it did not carry the picture element of the original, it did interlock perfectly with the neighboring pieces.

This is analogous to what we do when we automate, or when we apply computer technology to the improvement of some element of the work involved in manufacturing. We replace a human-powered, or a human-controlled, task with a machine-powered, or a data-controlled mechanism. But because we are confining our modernization to a small and localized area of the total manufacturing operation, the new mechanism must match the adjacent areas insofar as input, output, speed, and performance are concerned. If this were not so, there would be a disturbance caused by the change: either other parts of the manufacturing process would have to be changed also, or there would be a gap or overlap at the boundaries, or we would have to restructure our new mechanism.

Computer aided design, computer aided manufacturing, computer aided engineering, or any of the many other computer aided advances of the recent past are examples of this piece-by-piece enhancement of the manufacturing technology. Each is the application of an improvement in some particular portion of the total process. I have called these local fixes "band-aid" improvements. Each is a very real improvement, fully justified, and achievable providing that it fits into the environment with a minimum of disturbance to neighboring bits of the technology.

Despite the complexity and diversity of manufacturing technology, someone somewhere, and often many people, have succeeded in applying computer aided technology – or the mechanical equivalent, automation – to every single bit of the manufacturing world. All these bits are available to others. What more can we ask? If we persist long enough, we may automate or computerize our entire technology. (Common sense prevents us from trying to change all parts of the technology simultaneously; that is guaranteed to produce chaos.)

What we have done is to upgrade each of the atoms of the complex without improving the total structure. In spite of the crazy quilt of bandaids that cover the art, none of them cooperate with the others any better than did the human predecessors. They all use data; why don't they communicate? Because they were not designed to communicate! They were designed to communicate across borders that were originally determined and set by people-to-people data exchange modes. We have been replacing old bits of the jigsaw puzzle with glossy new, computerized and automated bits, but we have not changed the original pattern.

We all know why the *jigsaw puzzle* bits have the complicated interlocking pattern of edge boundaries – it is because they were designed to connect perfectly with the adjoining pieces, and only with those specific pieces. Getting all of them assembled, one to another, so as to form the whole was the heart of the puzzle. The pattern of manufacturing is much the same. It is composed of many small functions; the total task is divided amongst them, in accordance with a scheme that makes all the functions work together to produce the desired end result.

Why is the organization of manufacturing seemingly so complex? Years ago, when all of manufacturing was the application of human muscle power, guided by human brainpower, the organization of an enterprise was governed by the people available to do the work. When one had defined the types of people required and their relationship to one another, one had also defined the structure of the manufacturing process. The tasks were allocated in bits of a size to fit the capabilities of those available to do the work. Because people don't come in standard-sized abilities, the organizational "pattern was changed to fit the available cloth." As mechanization, and then automation and computerization followed, the basic pattern remained more or less the same, even though the communication between the parts was no longer human.

The time is now upon us to change this. We should organize the work of manufacturing to conform to the functions of manufacturing, not to the people or the tools used in manufacturing. Then we can expect the various bits to which we have applied computer technology to communicate and to cooperate wth one another efficiently. The jigsaw puzzle pattern can be replaced with a rational pattern, dictated by the fundamental, intrinsic patterns of the science of manufacturing.

My studies have convinced me that the structure of the science of manufacturing is the same, whether one is making airplanes, carpets, computers, canned soup, automobiles, paper clips, electric motors, washing machines, typewriters, lead pencils, newspapers, pianos, clocks, shoes, or television sets. The same basic functions are performed, the same basic managerial controls must be exercised. Furthermore, I believe that these functions and controls will be invariant over time, so that regardless of when or in what stage of technological development one is working, the same principles and understanding will apply. Many a manufacturer has told me, "We are different; our problems differ from those of any other enterprise." It always turns out that the differences are real but only superficial; the fundamental structure of manufacturing is and always will be a constant.

What is the structure of the science of manufacturing? When we know that, we can plan the reintegration of the many and curious bits with which we are now familiar. We can expect the flow of data and materials among the several functions to be continuous and effective. We can reduce the management of the acts of manufacturing to a similarly rationalized structure. We can expect the flow of command and report between the management and the execution functions to be smooth and efficient.

When we understand our profession as a science, we can subject it to well-known methods of analysis. Having analyzed, we can predict; we can determine what the basic parameters are and how they should be measured. When we can measure, we can control; and when we can control, we can succeed!

I believe that this concept of manufacturing technology will be the source of the greatest progress in the next 20 years. Step one of this task is to understand manufacturing as a science. Step two is to communicate this understanding to everyone involved in manufacturing— top level managers, marketing managers, research and development managers, production managers, middle managers of all sorts, supervisors and foremen, engineers, machine operators, apprentices, legal counsel, comptrollers and accountants, personnel directors and data processing managers. It should be taught in schools preparing students for entry into all phases of the industry. It should be learned by those serving the industry as vendors, and most certainly should be understood by those in government who must deal with the manufacturing industries. This book is devoted to that understanding.

Joseph Harrington, Jr.

# Contents

# List of Illustrations

# List of Blocks in the Model Diagrams

# UNDERSTANDING THE MANUFACTURING PROCESS

# Chapter 1

# Introduction

Manufacturing is evolving from an art or a trade into a science, an important one. A quarter of the population is involved in some form of manufacturing activity, and the rest of the population benefits from the products. When manufacturing was still an art, or rather a collection of very different arts, each had its own unique technology. We now see manufacturing as a science whose fundamentals are independent of what is being made, or when it is being made. It has a structure that is the key to understanding the science, and understanding is the key to profitable application of the science. Most important, this structure is invariant, so that once understood, the knowledge may be applied to any of the many technologies.

Learning is, in the ultimate analysis, the transfer of experience gained in the past to problems of the present. When we understand the scientific principles that govern manufacturing we can profitably apply that knowledge in many ways to the problems of today's highly integrated manufacturing technology. We can expedite the production of goods; we can adapt designs to changing needs more readily; we can increase the profitability of the enterprise; we can conserve materials and labor. The role of manufacturing in the economy, and in society, is changing with the times. With such a scientific understanding of the subject, we can analyze the effects of proposed changes and select optimum courses of action.

This understanding is essential for everyone involved, whether an old timer or a novice, whether a top level manager or an apprentice worker; it is essential for those in marketing and in production, for those acting as legal counsel or financial managers, for those in personnel or in data processing. It is most certainly essential for students preparing to enter the field of manufacturing in any of its phases.

It is the objective of this book to explain this structure of the science of manufacturing, and to show how it works and how to use it. Because the science described is generic to all the many various forms of manufacture, it may be applied profitably to any manufacturing enterprise.

## MANUFACTURING DEFINED

Manufacturing is the conversion of naturally occurring raw materials into desired end-products. The word derives from two Latin word roots meaning "hand" and "make" – almost literally "handmaking." In early civilizations products were indeed hand made; human muscle power and mental control moved crude tools over materials gathered by hand. Today, few products are made by human, or even animal muscle power. Other sources furnish the power, but humans still conceive the products and guide the operations of production.

In the broadest sense, manufacturing begins with the acquisition of raw materials, and extends throughout the whole gamut of activities of production to the distribution and, if necessary, the maintenance of the end-products. It is convenient to think of this sequence as having three parts – extraction of the naturally occurring materials from the environment, culling, and concentration; the conversion of these materials into a specialized form in bulk; and the conversion of bits of the bulk materials into discrete parts which, when assembled with other parts, constitute the desired end-products. For example, iron ore is dug from the ground, cleaned, and concentrated; it is smelted and converted to steel which is rolled into bars and sheets; pieces of steel are machined into pistons, connecting rods, and crankshafts which are assembled into engines.

In the first step, to take other examples, trees may be cut, limbs and roots removed, and the trunks taken to a sawmill; or aluminum ore (bauxite) may be dug up, cleaned, and shipped to an aluminum refinery; or cotton may be grown, picked, ginned, and shipped to a cotton mill;

or oil and gas may be pumped from a well, separated, and piped to an oil refinery.

In the second step, logs may be cut into boards or lumber, seasoned, sorted, or milled into plywood; or the bauxite may be smelted, refined, alloyed, and rolled into aluminum sheets, slabs, or drawn into wire or tubing; or bales of cotton may be carded, spun into thread, woven or knitted into fabric, and dyed or printed; or oil may be refined, fractionated and used as bulk fuels, or the gas may be chemically combined with other elements to produce the monomers for plastics.

In the third step, a factory may acquire a variety of bulk materials from the second step and shape individual pieces which are assembled into end-products, tested, and distributed to the end users. These end-products may be infinitely diverse – ships, airplanes, automobiles, telephones, radios, typewriters, computers, clothing, carpets, shoes, lights, cameras, clocks, furniture, and toys. The variety of consumer goods is seemingly endless. Some products are produced in large quantities – paper clips, for example; other products are produced in smaller numbers – fine automobiles, for example. Similarly, the end-products may be machinery to be used in making other products – printing presses, looms, metal working machines of all sorts, conveyors, retorts – again, the list of capital goods is seemingly endless in its variety.

It is not necessary that the ultimate product of this long sequence be either essential to life, or even useful in our daily affairs. Products may serve only to satisfy our aesthetic senses – fine art of all sorts, for example; or fine foods. If a product is desired, and is physically created by a sequence of acts, it is in the ultimate sense a manufactured product.

This book must confine itself to a manageable size and a limited field. It will concentrate on the third step of the three outlined above. This activity has become known as the "discrete parts industry" to distinguish it from the second step activities which generally produce bulk quantities of materials – feet of lumber, yards of cloth, rolls of steel strip; or pounds of molding compound.

How shall we define the word "manufacturing," which we shall be using so frequently? Word usage in this field is as diverse as the segments of the field, if not more so. Individual companies in a single segment may give the same word quite different meanings. The word manufacturing itself is a good example: in some instances it refers to everything the company does; in others it refers to everything except marketing; in still others it refers only to the fabrication department and excludes product design as well as marketing.

In this book, manufacturing will encompass the entire gamut of activities from product concept to maintenance of past products in the field, and everything in between. It will include product conception, product design, manufacturing engineering, fabrication of parts, assembly, test, distribution, and support. It will include all of the managerial functions necessary to integrate and operate the activity reliably, profitably, and in a timely manner.

In short, we will treat manufacturing as an indivisible, monolithic activity, incredibly diverse and complex in its fine detail. The many parts are inextricably interdependent and interconnected, so that no part may be safely separated from the rest and treated in isolation, without an adverse impact on the remainder, and thus on the whole.

This book is written to expound this complex structure, to trace its origin, to explore the interaction of the parts, and to understand how to measure, and therefore control, manufacturing. So understood, manufacturing will advance from an art to a science. This book is written, not to discuss how to perform one part of the task, but how to understand and manage the whole, or how to participate more intelligently in some part of the whole.


## THE HISTORY OF MANUFACTURING

That manufacturing is indeed a monolithic function can be seen in its origins. Anthropologists suggest that one of the things that distinguished early man from other forms of life was the opposable thumb, which made it possible for a man to grasp and manipulate a tool. Early tools were probably no more than a sharp edged rock fragment or a conveniently shaped stick. With these tools and his muscles, early man was able to skin and cut game for food and for leather clothing, to erect shelters, and to defend himself.

The Bronze and later the Iron Age introduced metals for tools — swords and spear points, axes and saws, planes and chisels, needles, pots and pans, and many other tools. Wood working and stone working skills evolved, and civilization moved forward. At the time of the Roman Empire, manufacture was well advanced in ship building, weaving, metal working, housing, leather working, jewelry, and the fine arts. Tools were refined, but motive power was still largely human. Animals, water, or wind powered the grist and saw mills.

So it remained to the end of the eighteenth century. Tools and skills were progressively refined, and some of the products were exquisite works of art or precision mechanisms. However manufacture was the province of a master craftsman. He was a gun smith or a weaver, a shoemaker or a cabinet maker, a printer or a cooper. He conceived his product, procured his materials, made his tools, and fashioned the product. He performed the entire function of manufacture.

From the earliest times, extra workers were added by the master craftsman to do the routine labor, and thus specialization of labor was introduced. As market sizes increased, marketing of mass-produced products was spun off to someone else, and the first of many divisive barriers within the fabric of manufacturing was introduced. When steam power arrived in the early 1800s, it became desirable to have larger factories, with many more people in an establishment. This, in turn, required several echelons in the hierarchy of supervision. While the owner or master retained overall control, department heads divided up the responsibilities and authorities over the several stages of the work, and more divisive barriers and compartments were created in the fabric of the organization.

In the late 1800s, the function of product development was split off as a separate unit with its own staff, separated by another organizational barrier from production. Research, as distinct from development, made its appearance in the early 1900s.

As volume of production grew, plants grew, and the number of echelons in the managerial hierarchy also grew. Multiplant companies appeared. The result of the last two centuries of evolution and expansion in manufacturing is a structure of great complexity – a fractionated and divided structure, with each little part trying to protect its turf and optimize its own performance. This of course has led to suboptimization of the whole. It has also led to communications barriers between the units at every level, horizontally and vertically.

The result is that today people see manufacturing, not as a single continuum, but as an aggregation of factions, and not always cooperative factions. There are differences in the training, education, and even in the cultural background of the people in these compartments. There has been a loss of perspective on the part of many. This is not good.

Basically these changes were the result of the division of the original one-man craftsman structure as the scope of the enterprise grew. Skills were specialized and divided. Authorities were specialized and divided. Now it is obvious that, if a function in manufacturing is divided into two parts, and the two together are to be the equivalent of the undivided whole, then the partial functions are interrelated. So, no matter how complex the product, nor how much is made, nor how many different products are to be produced, nor how intricate the process, manufacturing is still a monolithic, indivisible continuum.

In spite of the fact that manufacturing is, and always has been an indivisible continuum, the mode of operation has changed from time to time. In the earliest times, manufacturing was labor intensive. The cost of tools was small compared to the cost of the many man hours of labor that went into the simplest product. Cloth for clothing was, for example, produced by first spinning the wool or flax into thread – a hand operation using only hand tools. Next, the thread was woven into cloth on a loom which was a simple device available in many homes. The weaving operation required many hours of repetitive labor as the shuttle was thrown from side to side of the warp.

This example could be multiplied over and over. The key element was the labor expended. Manufacturing was indeed labor intensive.

When, as noted above, power was introduced to manufacturing, it was possible to introduce machines to perform the work formerly done manually. The power plant was expensive, and the machines which used the power and performed the work were expensive. It is obvious that there was no need for power until there were machines that could use it, and no incentive to build machines until there was power to operate them. They came together, in a slow step-by-step evolutionary mode. For example, in shoemaking, soles for many centuries had been attached to the shoe by driving wooden pegs through the outersole and the innersole. An apprentice whittled the wooden pegs by hand, while the master drove them into the shoe. A peg driving machine was an obvious possibility, but it was difficult to design a machine that could drive hand-whittled pegs. However when someone invented a peg-making machine (about 1850) that made uniform pegs, it was not long before someone else developed a powered peg-driving machine.

Throughout industry, machine followed machine, and in the course of the nineteenth century, manufacturing evolved into a process using many machines and their related power sources. All of these machines and engines were expensive. The key element in manufacturing became the capital equipment. Manufacturing was then a capital intensive industry. It is still largely so today.

We are now, in the 1980s, entering a third and new mode of manufacturing operation. It is characterized by the fact that every one of the many acts of manufacture, and every bit of the managerial control of those acts, can be represented by data. Data are generated, transformed, and transmitted. In the ultimate analysis, all of manufacturing may be seen as a continuum of data processing. It provides the one base to which all the parts of the process may be related, the one thread which ties all the parts together.

This concept of the continuum of data processing has become apparent concurrently with the evolution of powerful data processing mechanisms. Tremendous strides have been made in the past two decades in this art. It is hard to say whether the need to handle data drove the development of the tools to do so, or whether the expanding capabilities of the tools led to an appreciation of, and therefore, the use of the data handling capability. However, just as machines became an extension of man's physical capabilities, the computer will become an extension of man's mental capabilities.

It is also interesting to note that the emphasis on data has caused the reintegration of manufacturing. After the fractionation and compartmentation of manufacturing which occurred during the capital intensive mode, the new data intensive mode is going to witness the reintegration of the science. Manufacturing will become data intensive.

This book approaches manufacturing from this reintegrated point of view. It approaches manufacturing as a basic problem of making a desired end-product from bulk raw materials. We will approach it – not on the basis of how it has been done by others in the past – but by determining what the basic elements and functions are which we would create if we had no precedents, but were possessed of all the other skills and knowledge which we have today.

To put this concept in a different way: we will develop a formalized system of functions and data transfers applicable in principle to any and all manufacturing, rather than a description of the system used to produce some specific product. We will thus develop a model that may be used to analyze and evaluate any given manufacturing system. Using the understanding of the structure of manufacturing presented by the model, we can transfer, from one system to another, the lessons of our past experience. The art of manufacturing will become the science of manufacturing.

Because manufacturing is a science, it is subject to analytical procedures that can identify the rules which govern its behavior. We can predict the effects of changes in conditions. Understanding the science, we can correctly measure the performance of manufacture. If we can measure, we can control our performance. When we can control, then we can succeed. This understanding of manufacturing is the objective of this book.

## THE DIVERSITY OF MANUFACTURING

The variations in the circumstances of manufacture are virtually unlimited. They vary because of

The kind of product being made;
The technological environment of the product – mechanical, electrical, biological, chemical, etc.;

The manufacturing technologies available;

The managerial and operator skills available;

The level of experience with similar products;

The available choices of raw materials;

The demand for customization of individual products;

The pressure of time due to competition;

The number of units to be produced in total, and per batch;

The size of the company;

The geographical location of the plant;

and for many other reasons. (The order of listing above is not important.)

Another source of variety in the manufacturing industry arises in the stability or instability of production levels, and the frequency of changes in the product design. At one extreme we find, for example, the U.S. Mint at Denver. It has been making one cent pieces for decades. Except for a change in the year date on the coins, the design, the size, and the materials are fixed. At the other end of the scale are the "job shops," which manufacture parts for others. No two orders need to be the same, and orders may never repeat. Some plants, such as the Mint, may enjoy a uniform work load day after day. Others work only as orders are received, so that the rate of production may be hectic at one time and proceed at a snail's pace at other times.

Furthermore, each of the many varieties of manufacturing has its own jargon and word usage. For example, a common task is known in Plant A in a certain city as "Manufacturing Engineering"; at Plant B in the same city, the same job is called "Methods Writing." A sister plant to Plant B, owned by the same company but located in another city, calls that same job "Process Engineering"! Or take the term "Master Schedule." If you have six representatives from six different firms in a room, you can find seven definitions for that common pair of words!

Now there is nothing wrong with having a local meaning for a word, so long as all who use it understand it the same way. It is only when trying to write a text such as this that one needs to find and stick with some standard meaning for each word. I have elected to observe the common usage of the metal working industry, particularly that part of the industry that makes machinery. I find that there are counterpart usages in other kinds of industry, and that those who manufacture cosmetics, TV dinners, dresses, tennis racquets, and magazines can make word-for-word translations from my language into theirs and still find that the principles set forth here will apply to their work. Effort has been made, as each term is introduced, to clearly define its meaning in this book. A Glossary is appended to collect these definitions in an easily accessible form.

In developing the structure of manufacturing we will proceed in a logical path from product concept to product delivery. It is important to note, however, that all steps in this chain of events are in progress concurrently when multiple product orders and various product designs are considered. Finally, it is important to remember that there will be the usual stream of mistakes to be corrected, modifications to the designs or to the production quantities, and changes in priorities. All of these have to be provided for and coped with in the course of events in even the best run factory.

In spite of all this diversity and variability, certain functions, certain decision points, and certain data handling processes will universally be present. These form the backbone of the structure of manufacturing which we shall develop and use.

## THE USE OF A MODEL OF MANUFACTURING

It will be helpful to our understanding if we can represent the structure of manufacturing with a model. There are several forms of models — linear or one-dimensional, graphical or two-dimensioal, and solid or three-dimensional. A written description of manufacturing would be a linear model, with the words of the description following one another in the sentence structure. This type of model has been used many times, but in view of the complexity of manufacturing, it becomes wordy and confusing.

A graphical model will serve us better. A graphical model with which we are all familiar is a road map. It represents a piece of the earth's surface, with cities and roads displayed for reference. It has many merits: it is printed on a reduced scale, so that we may scan an entire state at one glance. It omits a host of details which are not relevant to the map reader's needs. Most important of all, it shows, at one glance, the relative location between several cities and towns. The eye can see, and the mind can comprehend, these multiple simultaneous relationships.

While three-dimensional models are most instructive, they are not suitable for inclusion in a book.

A model is not the real world — it is just a simplified representation of it — but it grows naturally out of the real world. For example, just suppose that you have as a hobby the making of fine furniture, and you perceive the opportunity to turn your skill into a profitable activity.

Figure 1.1 (a) represents your hobby activity. You will then add another activity – the selling of the furniture – as shown in Figure 1.1 (b). However you will also find that you are exercising managerial control over the whole enterprise, establishing policies and programs, obtaining resources, and managing people. You will also be using the help of peo-

(a)

(b)
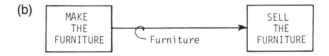

(c)
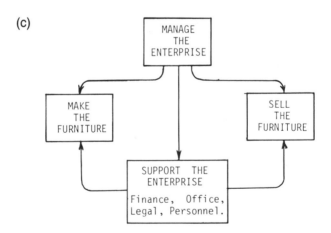

**Figure 1.1.** Development of a model. (a) The activity, when it was a hobby, had only one function. (b) The activity added the marketing function to become an enterprise. (c) Diagram of the complete enterprise.

ple with other talents – lawyers, accountants, etc. As a result, there will be four kinds of activities going on, as shown in Figure 1.1 (c). All four of these functions are essential to a manufacturing enterprise. Now we have built a model of your furniture business.

Just such a graphical model for manufacturing will be used, but employing a much more highly formalized modeling technique.

The modeling system is based on the Structural Analysis and Design Technique System (SADT) developed by Douglas T. Ross of the SofTech Corporation, Waltham, MA, and adopted and used by the U.S. Air Force Computer Aided Manufacturing Project, and used again in the U.S. Air Force Integrated Computer Aided Manufacturing Project, widely known as ICAM. This type of modeling system is fast becoming the common parlance of authors and workers in the computer integrated manufacturing world. It is known by an acronym: $IDEF_0$, which stands for ICAM Definition system. The subscript "0" indicates that it is a model of manufacturing functions and data flow. (There are other IDEF models: one with a subscript 1, which deals with the nature of the data handled; and one with the subscript 2, which treats the dynamic behavior of a manufacturing system.)

The purpose of the graphical model is to define and display the fundamental relationships of the many component functions which work together to accomplish manufacturing. The connections between the functions represent the flow of materials or information between the functions. How do we deal with the complexity of manufacturing? To quote Mr. Ross:

> The human mind can understand any amount of complexity – as long as it is presented in easy-to-grasp small chunks – that are structured together to make the whole.

In building such a model, we first present the "whole" and then examine the component "small chunks." These may each be successively expanded or "decomposed" to show more and more detail. Thus the model is concise and yet able to express complex relationships. It may be examined in detail, or seen as a total pattern.

The rules of modeling are set forth in texts, but are briefed in Appendix A. The modeling technology, and material from the $IDEF_0$ model as developed by ICAM, are used here with permission of SofTech and ICAM. It is suggested that the reader look over this Appendix before going further in the text, because we are about to start using the modeling technology.

Figure 1.2 represents the "whole." The box contains the function "Conduct a Manufacturing Enterprise." The activity is controlled by the

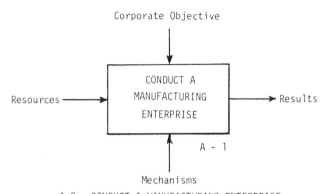

A-2    CONDUCT A MANUFACTURING ENTERPRISE

**Figure 1.2.** Highest level diagram in the hierarchy of diagrams that model manufacturing. The "A-2" in the diagram's title line is the Node Index number. The "A-1" under the lower right-hand corner of the box shows that this box will be decomposed and displayed in Diagram A-1. It may be seen in Figure 2.1.

statement of the corporate objectives. The function consumes resources, and produces results. (One hopes that the results produced satisfy the corporate objectives!)

Figure 2.1 shows the first expansion of the "whole" shown in Figure 1.2. It carries the same title, has the same control, the same output, and the same input as Figure 1.2, with the exception that the input has been expanded to include, not only the input funds, but also some of the items procured with the use of those funds. We will return to a discussion of Figure 2.1 in Chapter 2.

## OVERVIEW OF THE BOOK

Chapters 2 and 3 discuss the context within which manufacturing exists in a business enterprise, and the structure of such an enterprise. We found there are four major functions involved in the conduct of a manufacturing enterprise – management, manufacture, marketing, and support.

This book is directed to the details of the second of these four principal functions – the making of the product. This, too, has four major component functions contained within that simple concept – "make the product." The product or products have to be developed or designed. Next, the designs will govern the conversion of raw materials into the actual products. The products must be supported in the field. Finally, this whole manufacturing function must be supported and managed.

Chaptetrs 4, 5, 6, and 13 discuss these four major divisions of the manufacturing world in detail.

Chapter 6 deals with the production function, showing that it has six subfunctions. Owing to the complexity of this function, each of these subfunctions is discussed in separate chapters: Chapters 7, 8, 9, 10, 11, and 12. Chapter 12 deals at some length with the actual conversion of raw materials into finished products. It is in these chapters that we will be using our model to good advantage.

Having dissected the manufacturing function and analyzed it in great detail, we will reassemble the parts in Chapter 14 and look again at the totality of the manufacturing function that we first examined in Chapter 3.

Chapter 15 turns from the many functions and subfunctions of manufacturing to the connections between them, the arrows on the diagrams. It will examine the mechanisms of transfer of data from one function to another, the tremendous recent strides in our data handling technology, and the impact of that changing technology on the whole field of manufacturing.

It is the author's hope that this book will aid many different kinds of people in applying their understanding of the science of manufacture in dealing with their part of the total activity.

A Glossary has been included to define the precise meaning of the many terms used in very specific meanings on the model diagrams and in the text.

# Chapter 2

# The Context in Which Manufacturing Exists

A manufacturing enterprise is distinguished from all the many other forms of enterprises by its focus on producing goods, as contrasted to such other forms as service, transportation, communication, agriculture, education, banking, or government. We will direct our attention to enterprises which are devoted to manufacturing. It is possible, of course, to find a manufacturing activity embedded in one of the other types of enterprises. For example, a railroad might elect to make its own locomotives, or the Department of Defense might elect to make its own aircraft. In such cases we will assume that the manufacturing part is sufficiently different from the other parts that it may be considered by itself.

Even in an enterprise that is principally manufacturing, that function is performed in conjunction with other functions which together make up the total enterprise. It is important to understand this total corporate environment in which the manufacturing function exists.

The term "enterprise" may designate a single individual, a partnership, a company, a corporation, or even an arm of the government. All of these various sized enterprises may conduct manufacturing, but while the motivation and the control may be quite different, the implementation is similar enough in all cases so that our discussion will be applicable to any and all of them. To simplify the presentation, we will speak in terms of a corporate enterprise, rather than an individual or a government agency. This will gain for us some brevity and lucidity in language without in any way losing generality. The reader may easily translate the terminology into any other environment.

By assuming this corporate context, we also take cognizance of the fact that most manufacturing enterprises are on-going activities. We make mention of this because a discussion of manufacturing is most easily understood when presented as a serial description of all the various stages that a product traverses during its life cycle. In reality, in a given enterprise there will at all times be a variety of products in a variety of stages of their life cycles. There may also be repeat orders for production of any one of various products, in different steps in their production cycles, concurrently wending their way through the plant, and all contending for funds, machinery, and attention.

Figure 1.2, introduced in Chapter 1, is a representation of a manufacturing enterprise. Here we will examine the corporate context within which the manufacturing function exists, before getting into a discussion of manufacturing itself.

## THE FOUR PARTS OF A MANUFACTURING ENTERPRISE

Manufacturing-oriented enterprises perform four major functions, which together comprise the total enterprise. We found these four parts when we were discussing the model of a hobby that matured into a business, Figure 1.1. Quite obviously, one function is making the product; another is marketing the product. Manufacturing and marketing are two very different and clearly distinguishable activities. They require different facilities and skills, and use quite different kinds of people. Manufacture is an internally focused activity; marketing is an externally focused activity. In addition to these two major functions there is a very necessary managerial function, which again is clearly differentiated from the other two, and subtends both in its scope. Finally, there is a group of activities which are neither manufacturing nor marketing, and are not managerial. These are tasks which require expertise in the legal, financial, personnel, and data processing fields, and which support the more technical activities in the other functions. This fourth function is called the support function. Remember that all of these four are very closely interactive; no one can exist without the others.

The concept of the corporate context is important. Figure 2.1 is the expansion of Figure 1.2, showing the four major functions and the flow of data between them. Let us now examine in more detail what the functions are and what the people who staff them do.

### Managing

The staff plans how it will meet the corporate objectives. What kinds of products and services will they offer to what specific segments of the market place? They adopt corporate strategies, policies, and programs, and provide the necessary resources to carry them out. They issue the appropriate directives, and administer their execution. For example:

> We will concentrate our marketing in one area this year.
> We will lease rather than sell whenever possible.
> Model 4 should be ready to sell by the first of next year.
> A block of preferred stock will be sold next month.
> Division reports are due on the first of each month.

Box 1 of Figure 2.1 shows this set of activities as the function "Manage Enterprise." The entire enterprise operates under the control of the corporate objectives, but they are shown as a control for the "Manage Enterprise" function, which then passes them on down to the rest of the managed hierarchy. In addition, this function is controlled by information received from within concerning the market place, new product concepts, and the status of work in process. The input to this function is the financial resources of the corporation, which are also passed on to other functions in the hierarchy.

There are three outputs shown. First is the control exercised by the managers – the extension of their authority and mandate to manufacture, to market, and to support. Second is the communication of the policies and programs generated by the corporate managers – product definitions, marketing objectives, and support requirements. Third is the allocation of funds essential to the other functions to perform their tasks. There are two mechanisms shown. First, labeled "Experience, Facilities, Equipment, Tooling, Personnel, and Technology," is inherent in any on-going business. Second is the support of the financial, legal, data processing, and personnel talent resident in the support function, Box 4.

### Manufacturing

The staff develops and produces the corporation's product in a continuing operation, so that the products will keep pace with the state of the art of the buyers of the products, and so that the corporation's own pro-

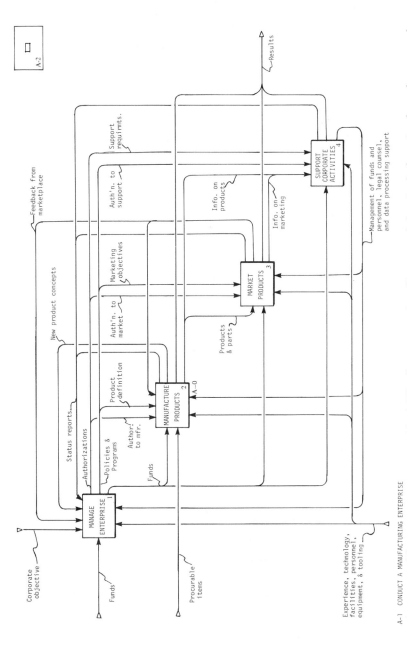

A-1 CONDUCT A MANUFACTURING ENTERPRISE

**Figure 2.1.** The corporate context (A-1). The first expansion diagram of the single function box shown on the A-2 diagram (Figure 1.2). It has the same title, inputs, controls, outputs, and mechanisms as does its parent, but reveals the four-part structure of the function "Conduct a Manufacturing Enterprise." Each box carries an identifying number in its lower right-hand interior corner. The "A-0" under box 2 indicates that it will be further decomposed in the next lower level diagram A 0 (Figure 3.2). Boxes 1, 3, and 4 will not be further decomposed.

17

duction methods will keep pace with the state of the art of manufacture. For example, suppose that our corporation makes shoemaking machinery. They will strive to make machines suited to produce the kinds and styles of shoes coming into popular demand. They will also strive to use the most modern metal working technology to shape the parts of their shoemaking machinery. In addition to the development department, there will be the production department, with machinery and skills necessary to convert raw materials into finished machines, test them, and support them in the field. There will be units to procure the raw materials, and units to design production procedures and provide the tools and equipment needed.

Box 2 of Figure 2.1 shows the function "Manufacture Product." This function covers all the activities which we have defined as included in the meaning of the word "Manufacturing" as we shall use it, and to which we will devote the majority of our subsequent attention. This function operates under the control of the authority and mandate to manufacture, which comes from the management function in Box 1. It is also constrained to make the products defined by Box 1 managers. The inputs are two – the funds allocated to the function by Box 1, and procurable items such as raw materials or finished components purchased from others. These inputs may be incorporated in the products, or utilized internally in the acts of making the products.

Again, there are three outputs, the principal one being the products and parts manufactured and turned over to marketing in Box 3. Information concerning the products, such as patentable inventions, product liability hazards, or personnel problems, are turned over to the support function, Box 4. Another output is a by-product of the research and development department – concepts for new products or new capabilities not included in the authorized product definitions. This output gets fed back to the management function, Box 1 for their consideration and decision. The other output is an intangible one, the contribution of the manufacturing organization to the overall effort to meet the corporate objectives. And as in Box 1, there are two mechanisms shown, the experience backlog and the support from Box 4.

## Marketing

The staff markets the stream of products by informing their potential market of the products; by contracting for the delivery of their products; by delivering them, and by installing them where that is called for. Also they perceive the changing needs of the market place for new or improved products, in larger or smaller quantities. Advertising, selling, and servicing are component functions of marketing.

Box 3 of Figure 2.1 shows the function "Market Products." It, too, operates under the control emanating from Box 1, including the authority and mandate to sell the products, in accordance with the policies and programs established by the corporate managers, Box 1. The input to Box 3 is the stream of products and parts produced by manufacturing in Box 2. They also receive as an input a block of funds allocated to them by Box 1 to carry out their function. And again, the output is three-fold, in this case the very tangible financial income from the sale of products or services in pursuit of the corporate objective. They also output information to the support group, Box 4, such as notice of market conditions, competition, pricing structures, personnel problems, and the like. The third, and very important output, is the feedback from the marketplace of information essential to the managers in Box 1 in their function of setting policies and programs. Marketplace information is also fed back to the manufacturing function in Box 1, particularly to the development unit for their use in designing new products, and to effect engineering changes to make the product more serviceable. And as in the case of both Boxes 1 and 2 they employ the mechanism of the experience backlog and the support from the support staff, Box 4.

It might well be argued that Figure 2.1 should display the box for Marketing in advance of the box for Manufacturing. It is true that no initiative for developing a new product is taken until there has been a perceived need in the market place adequate to justify the cost of the effort to design, produce, sell, and distribute the product. Indeed, the Factory of the Future Conceptual Framework, under development by the Air Force Computer Aided Manufacturing Project as this book is being written, does show contact with the potential customer as the first element in their A-1 diagram. In the aerospace world this is quite appropriate. However, the arrangement of Figure 2.1 makes a rational flow of manufactured products from manufacturing to marketing to the field, suggesting that the manufacturing box should precede the marketing box. The feedback from the marketplace to the development function is provided in our diagram, and in practice it will be found that this process is a closed loop, in which information and products flow round and round. This is a good demonstration of the fact that a function diagram, such as we are using, does not imply a timed relationship between the functions.

## Supporting

The staff provides legal counsel, financial services (such as the accounting department, the comptroller, etc.), personnel management (such as employment, training, fringe benefit administration, union relations, etc.), and data processing facilities, used by all the other functions.

Box 4 of Figure 2.1 shows this as the "Support Corporate Activities" function. It, too, operates under the authority and mandate of Box 1 to provide the support services, in accordance with the established policies and programs. Its actions are also controlled by the input of information received from manufacturing, Box 2, and marketing, Box 3. Both will report regularly on financial status relative to the budgets; manufacturing will give them information on the products, and marketing will give them information on its activities. Support has one input, the share of the funds allocated to it by Box 1. Its outputs consist of the support services described above to Boxes 1, 2, and 3, plus the contribution of the function to the overall effort to meet the corporate objectives.

These four kinds of activities, working together, constitute the single entity which may be thought of as a manufacturing enterprise.

## THE CORPORATE OBJECTIVE

The objective of a manufacturing enterprise is to convert capital into a means of production, and then to convert the resulting production into revenue to replace the used capital and result in a profit. Profit is returned to the owners either by increasing the asset value of the enterprise (capital) or by returning some of the profits to the owners as dividends.

Figure 1.1 shows the corporate objective to be the controlling element of the entire enterprise. Just what form would such an objective take? If the objective of a prototype corporation could be compressed into one sentence, it might read like this:

> The objective of the ABC Corporation is to earn a return on the investments of its stock holders and bond holders by developing, producing, and marketing equipment and supplies for the XYZ industry, for the foreseeable future.

This phraseology may seem to lay disproportionate emphasis on profit for the investors, but if it were not for the return on their investment, the capital so necessary for modern production would not be available. An enterprise must not only be legal and ethical, but profitable. If management forgets this factor, and the rate of return is not satisfactory, what is left of the capital will go elsewhere. In a capitalistic economy, profitability is not only acceptable, it is downright essential.

The phrase "for the foreseeable future" is most important. Survival depends on continued profitability, and that depends on the ability to meet both the current and the long term wants of the marketplace. This makes it essential to carry on research and development to meet the anticipated future. It also makes it imperative to plan and execute a long term enhancement program for facilities and technology. Of course, if the corporation is making hula hoops or some other product with a very short life expectancy, a different philosophy is in order.

The injunction for long term profitability has other benefits. It tends to protect the long term stability of the stream of profits; it gives workers a sense of job security; it assures buyers of the product of future support; and it tends to limit managerial risk taking. In short, it is the assurance of continuity of the corporation.

Addressing a specific marketplace – the XYZ industry, for example – indicates that activities are to be focused and not scattered, and that resources and efforts will be concentrated on that focal point. Specifying "equipment and supplies" in this case points to a full line of service, rather than a narrow line.

Drafting such a statement of objectives would be a very desirable exercise for any enterprise. It is not as easy as it might at first seem!

## THE GLOBAL VIEW OF THE CORPORATION

The objective discussed in the previous section emphasizes the importance of a "global" view of corporate activity when contemplating system-wide changes in organization or in operations, or both. The introduction of Computer Integrated Manufacturing (CIM) technology is a good example of such a system-wide change. Unlike a change in personnel assignment, or the replacement of one piece of capital equipment with another, a CIM system adoption has implications not heretofore considered in the usual acquisition justification exercises. The entire corporation financial structure is involved.

Acquisition of new machinery or facilities has conventionally been based on very narrowly focused justifications. The suggestion that an old machine A be replaced by a new machine B has originated within the organizational subdivision directly involved. The perception of the possibility arose from this narrow domain, and not from the broader corporate-wide perspective. The justification for the investment was based upon the before and after costs of the profit center's performance record. If other profit centers benefited by the change, well and good, but they did not contribute to the purchase of the new machine; and if they were harmed – let each unit defend its own interests! This was a parochial rather than a global view of the business. The presumption was that new machine B would pay for itself within 1 or 2 years; if not, there were better places to invest the funds.

This parochial approach is changing rapidly under the influence of a switch to the introduction of system-wide changes such as Computer Integrated Manufacturing. Such total systems concepts originate in sophisticated sources concerned with the well-being of the total corporation, and not within a single sub-unit. Their impact is company-wide rather than localized, and therefore the justification is based on the before and after costs to the whole company. The improvement in the return on investment comes more slowly because large changes take place more slowly. On the other hand, systems such as CIM will increase in their payback as time goes on, rather than declining in value as in the case of new machinery with use and obsolescence. Because of this company-wide involvement and because of the long time period of the commitment, the introduction of such global changes must involve top management in the justification process.

Whereas in the past the Chief Executive Officer (CEO) could rely upon his subordinates for the validity of the details justifying acquisitions of capital goods, without necessarily understanding the technology involved, acquisition of system-wide changes such as the adoption of CIM requires that the CEO fully comprehend the technology, and its implication for every nook and cranny of his organization. In general, the adoption of such global changes do not depend upon economic justifications, but upon strategic decisions, and therefore only the CEO is in a position to make the necessary decision and assume the necessary responsibility.

## THE MODEL OF MANUFACTURING AT THE ENTERPRISE LEVEL

In subsequent chapters of this book we will expand, or "decompose" the second box of Figure 2.1, labeled "Manufacture Products" and examine it in detail. It is not within the scope of this book to decompose Boxes 1, 3, and 4. However, a node tree, which is explained in Appendix A, is shown in Figure 2.2. This shows the component functions on the next lower level, but without showing the inputs, outputs, controls and mechanisms, and the flow of data between them.

Please note again that the discussion of functions in manufacturing is independent of the managerial and organizational structure of the enterprise. While organizations may follow the functional diagrams,

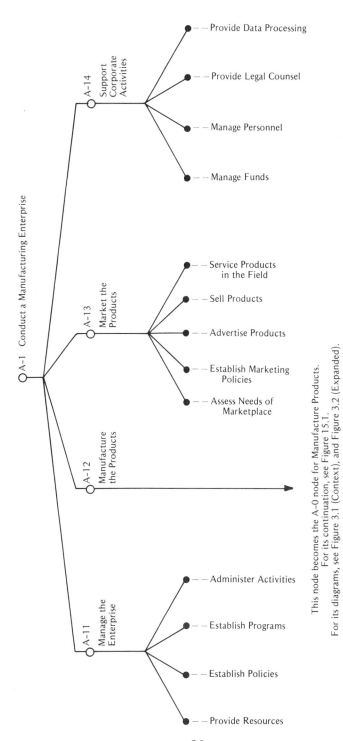

This node becomes the A–0 node for Manufacture Products.
For its continuation, see Figure 15.1.
For its diagrams, see Figure 3.1 (Context), and Figure 3.2 (Expanded).

**Figure 2.2.** NODE TREE for "Conduct a Manufacturing Enterprise." The second line of nodes correspond exactly to the function boxes in the A-1 diagram (Figure 2.1). Each, in turn, has subsidiary nodes. Only the A-12 node will be further decomposed, in Figure 15.1.

A–1  Conduct a Manufacturing Enterprise

A–14  Support Corporate Activities
— Provide Data Processing
— Provide Legal Counsel
— Manage Personnel
— Manage Funds

A–13  Market the Products
— Service Products in the Field
— Sell Products
— Advertise Products
— Establish Marketing Policies
— Assess Needs of Marketplace

A–12  Manufacture the Products

A–11  Manage the Enterprise
— Administer Activities
— Establish Programs
— Establish Policies
— Provide Resources

they do not always do so. Department names are not necessarily the same as functional words. One person or department may perform several functions; or one function may be performed by several persons or departments located in various parts of the company.

Organization charts show the hierarchical chain of command and the spheres of influence of individuals. However humans do not always come with the exact level of skill necessary to fill a given functional responsibility as shown on our diagrams. Each person is assigned as much authority as he can handle, which usually makes the organization chart depart from the functional structure. As he or she grows in capability, the organizational arrangement may be adapted to match.

It is expected that an appreciation of the functional relationships may in the future come to influence organizational structure. The reason is that the boundaries between functions are invariant, and do not change with time, whereas the boundaries between organizational domains rarely are permanent. The latter may be set by any number of things – competence and skills of the current domain managers, geographical dispersion, and tradition in the company – to name but a few.

## SUMMARY

An enterprise principally devoted to manufacturing has four major functional divisions which, working together, constitute the single entity: *Manufacturing*, which develops and makes the products; *Marketing*, which sells and services the products; *Management*, which orchestrates the other divisions; and *Support*, which provides specialized services to the others.

It is important to view Manufacturing in this corporate context, because it cannot operate except in conjunction with the other divisions. The corporation's objective will govern the operation of the enterprise. This global view of the corporation is particularly important when considering the introduction of company-wide systems such as Computer Integrated Manufacturing.

The balance of this book will concentrate on the Manufacturing function.

# Chapter 3

# The Manufacturing
# Function of an Enterprise

The manufacturing function of an enterprise is one of the four principal functions of the total enterprise, as described in Chapter 2, and as shown in Box 2 of Figure 2.1. It, in turn, is divided into four major subfunctions: the Management, Development, Production, and Support functions.

Quite obviously, if a company is to be in the manufacturing business, they must develop a product design, and then they must procure and convert raw materials into that product. These two major functions of the manufacturing division occur one after the other in the life cycle of any one product, but in most manufacturing enterprises with a number of different products or product lines, they are going on concurrently. Additionally because products in the hands of users need support, another function is needed to provide support for the products after they are turned over to Marketing, sold, and are in use in the field.

As on the higher level, there is the very necessary managerial function to coordinate, steer, and motivate the other three.

Figure 3.1 is the parent diagram for all of the Manufacture Product models. Following the IDEF$_0$ methodology, it is the exact equivalent of Box 2 on the higher level diagram, shown in Figure 2.1. It is shown in complete detail in Figure 3.1.

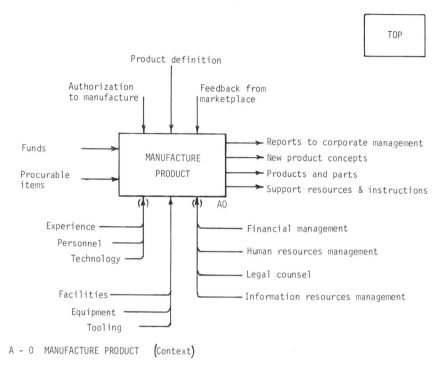

A - 0   MANUFACTURE PRODUCT   (Context)

**Figure 3.1.** A-0 MANUFACTURE PRODUCT. Before decomposition of Box 2 of the A-1 diagram (Figure 2.1), it is restated alone in what is called the "Top" or "Context" diagram of the succeeding hierarchy of diagrams. Note the parentheses around two arrowheads on the bottom of the box, showing that these mechanisms are implicitly passed on to lower levels of the diagrams, but to reduce the complexity, are not explicitly shown beyond (below) this point.

The things that control the manufacturing function are the mandate to conduct manufacturing in conformance with the product definition and specification received from the company overall management. One of the functions of the Manage Enterprise group (Box 1 of Figure 2.1) is to translate the strategic business plan of the enterprise into tactical and operational plans. The latter must be consistent with the corporate objective, but appear in a much more concrete form. They take the form of policies and programs. Upon the advice of the Marketing group, and with the concurrence of the total corporate management, they may elect to pursue the production and distribution of some new product. One element of that decision is a very clear definition of just exactly what that new product should be. The definition is given in terms, not of how the product is to be designed and built, but in terms of what it must offer to the buyer. (This last point is one that frequently generates friction between Manufacturing and Marketing: "Don't you tell me how to design the products!!" or "Don't tell me how to make them! I'll make them, you sell them!!") The definition also includes the intended rate of production, the probable ultimate market, and the desired date of initial introduction.

In the case of instructions to Manufacturing to undertake a project, it is necessary to authorize the manufacturing group to expend money and use recources in the execution of that directive. Without money and facilities, nothing would get done. However, along with the authority conferred, there goes the responsibility to perform on schedule, within the budget, and to product definition specifications.

Another of the functions of the corporate management is the provision and allocation of resources for the total activity. They appear as inputs to "Manufacture Product," Figure 3.1, in the form of the funds and the procurable items purchasable with some of the funds. The "Manufacture Product" function converts these inputs into the desired outputs – finished products, such as machines; parts which are sold separately for spare or repair use; support resources and instructions; such as manuals and training, and service tools; and that fortuitous output, the new product concepts and technical information generated in the research and development work. However the major output of Manufacturing is the product stream, delivered to Marketing to be sold or otherwise delivered to users.

Another output is the support that Manufacturing must give to Marketing to sustain the products in the field. Some consumer products, such as clothing, household articles, and the like, are not usually supported by the maker after they are sold; it is up to the owner to maintain them. However most capital goods, such as machinery, automobiles, or office machines usually must be supported in use by the maker. This means a supply of repair parts made available through distribution centers. It frequently means a service agent who can go to the user site, diagnose problems, render advice, and if necessary, make repairs. It certainly means the provision of user instruction and maintenance manuals.

Figure 3.1 also shows the mechanisms involved in the exercise of the function: facilities, such as buildings; equipment, such as machine tools; tooling, such as cutters, jigs, fixtures, dies, forms, and gauges; and the personnel to use these mechanisms. All of these are fairly tangible mechanisms, but there are two other classes of mechanisms essential to perform manufacturing. One is the backlog of experience and knowhow resident in the personnel and implicit in the machinery and tooling; the other is the knowledge of the technology that will be applied in the course of manufacture. An example of this might be the preferred method of arranging the cuts and bends in a progressive die for forming sheet metal parts, or the optimum feeds and speeds for cutting metals.

Note that two of the mechanisms are shown with parentheses around the tip of the arrow. That means that these mechanisms will pervade all the subsequent expansions of this parent box, and therefore will not be repeated over and over.

Figure 3.1, in turn, becomes the parent of the diagram shown in Figure 3.2. The latter diagram, however, shows the decomposition into the four functions discussed above, and the data flows between them. We will now discuss each of these four functions, with reference to the four boxes in Figure 3.2.

## THE MANUFACTURING MANAGEMENT FUNCTION

The manufacturing management function wields the driving intelligence and force or the whole manufacturing operation. It is focused on the task of steering the design, production, and support activities solely within the manufacturing function, and operates from the beginning to the end of each product's life cycle, and on all currently active products. Please note that manufacturing management differs from the corporate management function, which is concerned with the entire enterprise; we must be specific. Hence the stress on "manufacturing." Management occurs on all levels in a company. The *Wall Street Journal* reported (July 12, 1982) that the General Electric Company had 400,000 employees, 25,000 of whom had the word "Manager" in their title.

The manufacturing management function, shown in Box 1 of Figure 3.2, is allocated the necessary funds to perform the total manufacturing function. The input is shown as an arrow entering the box, but the arrow head is surrounded by parentheses. This means that in the decomposition of this figure in later diagrams the distribution of the funds is implicit in the directives issued to the other functions, and will not be explicitly mentioned again.

What the manager of manufacturing does distribute to the other functions in the department is a set of directives. They direct the Product Development function to develop and design products in accordance with the product definitions and requirements received from cor-

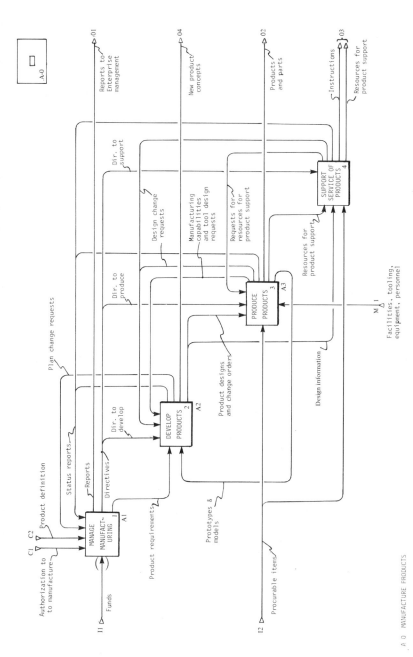

A-0

A 0  MANUFACTURE PRODUCTS

**Figure 3.2. Manufacture products (Expanded) (A 0).** Decomposition of the single parent box in the A 0 diagram (Figure 3.1). It has the same title, inputs, controls, outputs, and mechanisms as A 0, except that the two mechanism arrows with parentheses in A 0 do not now appear. Boxes 1, 2, and 3 have indicia below their lower right-hand corner indicating that they will be further decomposed in later diagrams: A 1 in Figure 4.1, A 2 in Figure 5.1; and A 3 in Figure 6.5.

29

porate management. They will include time schedules and budgets. They direct the Production function to produce a certain number of each product, on a time schedule and budget, in accordance with the designs received from the Product Development function; and they direct the Support Service of Products function to provide for spare and repair parts, operating and maintenance instructions, and the like.

Of course, manufacturing management renders regular status and financial reports to the corporate management and support functions. If for some legitimate reason Manufacturing cannot meet the product definitions, or the time schedules, or the budgets mandated by corporate management, early and clear representations to that effect are due back to corporate management. These feedbacks are shown in Figure 3.2 as one of the outputs from Box 1.

Included in this feedback is information to the legal advisors of patent possibilities, potential liability problems, labor relations, and the like. Also fed back are status reports on the course of the financial conduct of the manufacturing function. Very importantly, corporate management is apprised of developments arising in the research function which may offer opportunities for new product endeavors.

## THE PRODUCT DEVELOPMENT FUNCTION

Product designs are created and maintained by the development function, shown in Box 2 in Figure 3.2, and labeled "Develop Products." It is solely responsible for the designs, and has sole authority over them.

Development is controlled by a directive from manufacturing management to design a product, the specifications for which have been received from corporate management, together with an allocation of time and funds. This mandate carries with it the responsibility to maintain the product's design throughout the product's life. It is an unusual design which is not subject to change with time, but the development function is the only agency authorized to make changes in a design. The reasons for changes are manifold, as we shall see.

A variant on this pattern may occur when corporate management may decide that, rather than continue maintenance updates on a product design, the time has come for a complete redesign and the creation of what in effect is a new product.

Another variant on this pattern may occur in the case of the job shop, which makes parts to order for other enterprises. The parts have been designed by the prospective purchaser, so that the product design function may be only nominal, or possibly nonexistent in regard to those parts.

The development function operates on two bodies of past experience and engineering knowledge. The first is the technology of the industry which uses the product. If the company makes paper making machinery, the product designers must be fully familiar with the technology of paper manufacture; the company's research facility may in fact be one of the leading generators of that technology. Competing machine suppliers, the paper makers, and academia may also be generating new technology and amassing experience, and it is up to the product design function to keep abreast of the best of it.

The second is the technology of manufacturing the types of products offered by the company. In a metal working plant, this would mean the latest technologies for metal cutting and forming, quality control, assembly, and testing. It must be presumed that when a designer configures a part for a machine, he knows that it will be strong enough and stiff enough to function correctly; that it can be produced in a factory; that it can be assembled; and that, when assembled, it will work as it was designed to work. It is the designer's responsibility to make cost-effective designs.

Inasmuch as the development function is responsible for generating the product designs, it must retain sole control over them forever. If someone else is allowed to change the designs for any reason – good or bad – the product design function can no longer be held responsible for the performance of the product. The responsibility becomes divided, and trouble ensues. The rule in manufacturing is that the generator of data is solely responsible for it.

The outputs of the development function are product designs in complete detail, suitable if authorized for manufacture either in-house or by some other enterprise. Designs consist of layout drawings showing the total product, with its component parts in working relation to one another; of detail drawings, one for each individual part, so thorough that another person, at another time and place, can pick up the drawing and make the part so that it will fulfill the design requirements; and of a list of all the parts required for the assembly, arranged in hierarchical array showing the sequence of assembly and the number required. Details of parts may identify parts to be purchased complete rather than made from scratch. The list of parts is known as the Bill of Materials (BOM) in most industries. Accompanying the drawings and the bill of materials may be models, pictures, sample parts, or similar aids to understanding.

Another and very important output of the development function is, as mentioned above, research data and concepts for possible new products not previously contemplated by others.

Needless to say, the development function sends financial and status reports to the management function, plus statements of problems which may require modifications to budgets, schedules, or product definitions. Design information is also supplied to the support function so that it may prepare to support intelligently the products when they are in use in the field.

In order for the research and development work to go forward, experimental models and prototype devices are required. The production function, being equipped to make such things, is called upon to do so. Organizationally, the model shop may be separate from the production shop. It will probably be equipped with versatile machinery particularly suited to small lots and model quality levels. Nevertheless, the function of producing models is part of the Production function. In its operation it performs all of the subfunctions of production, has all of the managerial problems, and has all the procurement and resource supply problems as does the shop producing the finished products. Hence, the model and prototype construction is treated in this book as a production function.

## THE PRODUCTION FUNCTION

The production function converts procured materials into finished products.

The production function receives from the development function the product designs and the product design modifications, which are its principal input. They will control the conversion of the supply of materials into the principal output – the actual products. The supply of materials includes the procurable items – all the raw materials, finished items bought from others for incorporation in the products, tooling, and items acquired for conversion in-house into resources to be used in manufacture of the product. The first three of these are fairly self-explanatory, but the fourth may need this added word. In the metalworking factories it is usual for the plant to design and build many fixtures and devices used in the course of manufacture but not incorporated in the product.

The production function operates under the control of the manufacturing management, which gives it the directives to produce, along with an allocation of funds, resources, and time. As may be seen in Figure 3.2, the product designs are thought of as controls for production, rather than inputs, because they are not consumed and passed on with the products, but remain for subsequent production of more of the same products. It is the procureable materials which are consumed, converted and passed along as output.

The outputs of the production function are of course the finished parts and assembled products and the spare and repair parts, which are transferred to the responsibility of the Marketing function. Another output is the information for the training and maintenance manuals. Still another form of output is the model and prototype work done for and delivered to the development function for use in their experimental efforts. These latter outputs are usually made by the shop for the development labs, and may be built and rebuilt many times before they have completed their usefulness. Only the lessons learned from them get incorporated in the ultimate product stream delivered to the customers.

There are also intangible outputs from the Production function. Financial and status reports are sent to the management function, as might be expected, so that they may keep track of progress. Statements of the shop's manufacturing capabilities are sent to the designers so that they may focus their designs on configurations which the shop can make. If the designs cannot be made in the shop, the shop will either have to go outside to have the nonconforming parts manufactured, at an additional cost in time and money, or the shop will have to acquire new machinery capable of meeting the design requirements.

Inevitably during the course of manufacture there will arise the need or desire to have a design modified. A change in a part design, if acceptable to development, might facilitate manufacture, or conserve materials, or facilitate assembly or service, or save time in an otherwise tight schedule. Requests for design changes are referred back to the design function, and may come back down through the system as "engineering change notices." If for any reason a change is authorized, Development must not fail to change its master record to conform.

Production plans how each product is to be made, and how each part of each product is to be made; and how they are to be assembled and tested. This is a long and arduous process, but it is only done once for each part (until that part is changed by one of those engineering change notices!). Intertwined with this planning function is the job of making and administering production schedules and budgets.

Production receives, inspects, accepts, and stores incoming material; fabricates it into parts, inspects and stores them; assembles and tests the completed products; and ships them as directed. These functions are repeated many times – once for each part made, once for each assembly made. This repetition is in distinction to the planning function described above, which is done but once for each design.

It is worth noting here that in many environments, when the word "manufacturing" is used, it means what has just been described as the production function, and not the broader function as the term has been used in this book. Local usage is fine when used locally; in this book we will stick to the broader scope of the term's definition.

## THE SUPPORT OF PRODUCTS
## IN SERVICE FUNCTION

The fourth major component of the manufacturing function is the support of service for products in use in the field. Products do wear out, and sometimes break or are damaged. Repair parts must be made available to the product user, and sometimes over a very long time period and over a very wide geographical area. It is the responsibility of Marketing to carry out all the direct contacts with users, but Marketing's service agents must be equipped with a supply of appropriate parts, special devices needed to install the parts, and manuals for the operation and maintenance of the products. Representatives of Manufacturing's support function do not usually perform the field work, but they do support those who do.

This function, shown in Box 4 of Figure 3.2, operates under a directive to perform the function, received from the managers of manufacturing. It receives information from both development and production, which it converts into its advice and support output. It may request that production make spare or repair parts available. Additionally it may send back to development a request that a part design be modified to reduce breakage, to prolong wear, and thus reduce service calls, or to make the servicing easier when it is necessary. If Development concurs, they change the designs and issue an engineering change notice to production to change the parts.

## THE CONTINUITY
## OF MANUFACTURING

The complexity of manufacturing stems partly from the fact that all of the several steps in a product's life cycle, and all of the various stages in the conversion of raw materials into finished products, are inextricably interdependent and interconnected. The complexity may be appreciated when we examine the timing and frequency of these various steps and stages.

Designing is done once for each product, but the design is subject to revision from time to time as experience dictates.

Planning for manufacturing is done once for each product or each individual part to be made, and is subject to frequent revision. Production technology itself is subject to periodic improvement.

Production is done once for each product instance – each serial number of a machine or each separate part to be made. If there are a hundred parts in a batch, the production cycle is repeated a hundred times. When production is "to stock," meaning that the company is producing units to meet a forecast demand and is storing each batch of parts made until they are consumed, each work order entered in the shop constitutes a one time production of many parts. When production is "to order," meaning that a machine or some special part is made and will be sold when completed, each such order constitutes a one time production of the appropriate number of production cycles.

Each product design has its own time cycle and sequence with which it moves through the several design and planning steps.

Each part has a routing that it follows from procurement of the raw material through the various steps of manufacture, moving from machine to machine through the shop.

Each machine tool or other facility has its own finite capacity, dealing with one part at a time.

It is easy to see that, with a limited set of machines and operators, a limited number of managers, and a limited amount of time, there will be conflicts. Something will have to wait for something else, and when it waits, all its subsequent planned steps get delayed. Queuing is required, and priorities must be established on the basis of the optimization of the overall manufacturing operation. Such management has been described as a three-dimensional chess game, in which the planners try to get hundreds of individual parts made, each following its own unique route from machine to machine, using a limited number of machines and operators, and meeting a closely scheduled program of part needs at the assembly floor.

On top of this, unplanned delays occur. An operator is out sick, a machine is broken, a tool is missing, a material on order was not delivered, there is a power failure, and so on. As a result, schedules are in a constant state of change. In fact, a state of change is the only normal condition!

Clearly, the high degree of continuity and connectivity of manufacturing makes running such a function a complex management task.

## SUMMARY

The manufacturing function develops products and produces them, and it supports the products in the field. In addition to these three functions there is the necessary managerial function, making the four major subdivisions which comprise manufacturing.

Manufacturing develops product designs to meet the product definitions and specifications by conducting research, developing product configurations, and documenting the product designs.

Manufacturing procures raw materials and converts them into the parts of the product; it assembles the parts and tests the products before delivery to the users. A product may be manufactured once or many times.

Manufacturing supplies parts, instructions, and resources to enable the marketing unit to maintain the products in successful use in the field.

These three functions are integrated and directed by a managerial function, which directs and administers the total manufacturing operation.

# Chapter 4

# The Manufacturing
# Management Function

In Chapter 1 we discussed manufacturing in general, and adopted a definition for the purposes of this book. In Chapter 2 we discussed the corporate context in which manufacturing exists, and noted that it is but one of four major functions that go to make a complete corporation. In Chapter 3 we discussed the general structure of that manufacturing function and the four major functions which in turn comprise the manufacturing function. A glance at the node tree, Figure 2.2, will put all this in perspective.

Chapter 4 is the first of four chapters that will take up these four manufacturing functions: Management, Development, Production, and Support. Chapter 4 addresses the actions involved in managing a manufacturing activity. While we discussed this in chapter 3, we were there dealing in broad brush generalities. As pointed out in Appendix A, the higher level diagrams such as Figure 3.2, and the discussion that goes with them, contain a vast amount of implied information but little detail; the lower level diagrams and the discussion that turns around them, such as the ones we are about to generate, contain less information but display it in more detail. For a complete understanding of manufacturing we must investigate that detail.

The manufacturing management function wields the driving intelligence and force for the whole manufacturing operation. It plans for the conduct of its task; it schedules and budgets for each of the projects and products that are active; and it administers those plans and budgets. It is responsible for three other functions—development of the product designs, production of the products, and support for the products in the field. We will discuss development in detail in Chapter 5, Production in Chapters 6 through 12, and support in Chapter 13.

Let us assume for the purposes of this discussion that we are considering the management of an enterprise in being and in operation. (The managerial tasks of setting up a new business are somewhat different but will be an obvious variation of the tasks of a manager of a going concern.) The person charged with the management task will be fully aware of the product line, the staff and their capabilities, the plant and its equipment, and the current work load. Management of manufacturing is an on-going function.

Purely for the sake of clarity in explanation of this function, let us follow the manager's sequence of activities in starting up a new project —say, for example, a new objective established by corporate management to bring out a new product line. The manager's first task is planning.

## PLANNING FOR MANUFACTURE
## OF A PROJECT

When a new project is assigned to manufacturing, the manager will first seek to understand the task. How does it differ from the projects now in the mill? Is it a brand new concept, or is it merely a variation on a familiar theme? How big a task is the new project? Does the manager have all the necessary information with which to proceed? If not, who has it? This last question may have been answered in the corporate level discussions which led to the adoption of the new project by that group; doubtless the manager of manufacturing was a party to that decision, along with the manager of marketing from whom the original stimulus may have come.

The decision by corporate management should include a clear statement of the product definition and specifications. For example, if our enterprise is making high strength fasteners for the aircraft industry, the product definition should state the material, the finish, the tolerances, the configuration, and the performance criteria for the fastener.

It will indicate the potential market for the fasteners in the months ahead, and may hint at the ultimate life span of the product. On the other hand, suppose our enterprise is making turret lathes, and the new project is to devise an automatic loader for the lathes. If this is a brand new concept to the enterprise, there will obviously be extended development work involved. One of the common flaws in the definitions of such assignments is a misleading statement of the objective. It is liable to arrive, couched in the form of a machine concept. Preferably, it should be a clear statement of the function which the loader is to perform, phrased in such a fashion that the wording does not predetermine the configuration which will result. That should be left to the machine designer.

When the manager of manufacturing is confident of a clear, clean assignment, the planning process begins. First, the manager determines the extent and nature of the design and development work necessary, and the availability of competent design talent. Our fastener maker will look to see if the designers have the requisite technology at their command, or will they need to do some research? Will they fully understand the requirements? What past experience has the company had in this particular form of fastener manufacture? Is there an existing product in their catalog which can be used or adapted? Our lathe manufacturer will probably anticipate an intensive study period in which one or more of his engineers will catch up with the art of materials handling, robot loaders, and continuous part feeders. They will have to examine the range of the lathe product sizes which the new loader will have to handle, and the frequency with which it will have to function. In either case, our manager will consider whether or not he has the talent available to meet the requirements, and he will "guestimate" the length of time and the cost of the development task.

Next, the manager of manufacturing will determine the nature and extent of the production facilities required for the new project, and their availability in the time period in which he expects they will be needed. Obviously, the facilities are geared to the current product lines; how will this new project affect them? Is there unused capacity on the production machinery that the manager would be happy to see busily employed, or will more machine tools have to be acquired to carry the added load? Plant facilities and machine tools are expensive capital equipment, and are not bought lightly, nor are they quickly acquired. Justification and selection of capital equipment may take several months, and delivery and installation may take from a month to two or more years, depending on the length of the machine order backlog in the machine tool industry, and the size of the machinery ordered. If new equipment must be acquired to carry out the new project, then the sooner the acquisition is started the better. The alternative to acquiring new machinery is either to add another shift to the factory schedule or to farm out to a subcontractor the overflow work. If another shift is contemplated, then the availability of both operators and supervisors is a question. If subcontrating is contemplated, then the availability of appropriate firms and the added cost of their profit on the work must be taken into consideration.

The manager of manufacturing will assess all these production alternatives and again "guestimate" the cost and the time entailed by the new project. In a similar manner the manager will estimate the requirements which the new project will add to the product support function.

Next, the manager of manufacturing must compare the time and cost estimates for the project with the constraints of time schedules and budgets allocated to him by the corporate management in its mandate to him to manufacture. If the estimated schedules and budgets are within bounds, then the manager feels it is safe to proceed; if not, the manager should immediately go back to corporate management and negotiate a resolution of the conflict. Let us assume that any problems are resolved and that the manager may proceed to the next step in the managerial functions.

The manager creates a detailed plan of procedure. All the manager's subordinates are apprised of the plan, formal orders are distributed, and the project goes forward.

Once the project is launched, the function of the management of manufacturing becomes one of monitoring activities, receiving reports of status, and rendering reports to corporate management. In the real world it is not quite that simple. Unforeseen problems develop, and the managerial function is responsible for resolving them by modifying the plan to work around the problems, but still within the schedule and budget constraints. This function is the one referred to as administration. When problems cannot be resolved within the function's boundaries, then the manager must ask the corporate management for advice and possibly a change in their mandate to manufacture.

It is appropriate at this point to refer to Figure 4.1, entitled "Manage Manufacturing." This is an expansion of Box 1 in Figure 3.2, which carries the same title. Figure 4.1 shows the breakdown of the management function into three sub-functions of which the first is "Plan Projects," Box 1. That is what we have been discussing in this section. In the following sections we will explore the other two boxes and the functions they represent.

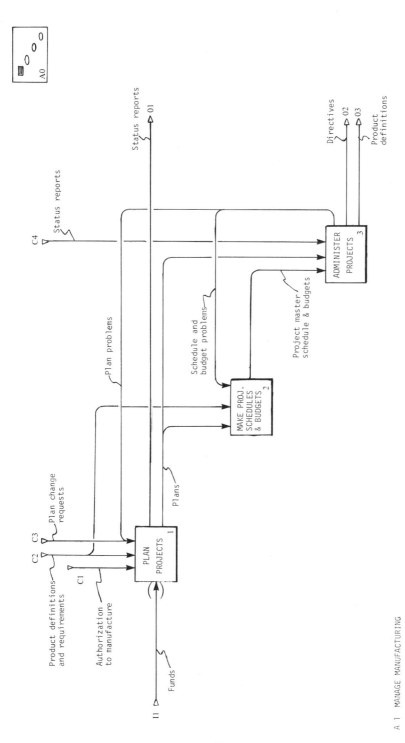

A 1  MANAGE MANUFACTURING

**Figure 4.1. Manage manufacturing (A 1).** Decomposition of the "Manage Manufacturing" box (Box 1) of diagram A-O (Figure 3.2). The small symbol in the upper right-hand corner of these diagrams suggest the layout of the parent diagram, with the solid color mark designating which of those boxes was the parent of this diagram.

41

## MAKE PROJECT SCHEDULES
## AND BUDGETS

For each new project, or for each individual production order entered, a schedule must be created and budgets established. These will become the milestones against which the manager will measure the performance of his organization and the achievement of his objectives. The general scheme of the manufacturing plan was described under Planning for Manufacture in the preceding section. It is necessary to refine this plan by going through the process in much greater detail, refining the plan in conference with the managers of development, production, and support. At each iteration, the total cost and time estimated are checked against the allowed costs and times.

The refinement process is largely one of breaking down the procedures into smaller bits and considering each by itself and as a part of the whole project, then combining the parts to form the whole. A milestone may be set at the conclusion of each of the smaller parts.

Suppose, for the sake of clarity in explanation, that the project calls for the development of a new machine to produce "widgets." Some research seems necessary if this model is to be superior to older and to competing models of widget machines. Thereafter there will have to be some conceptual design work, and a prototype machine built to test the newly researched technology and the newly conceived design. When the prototype is accepted, there will be a preliminary design made, followed by more testing. When this test is passed, the design will be reworked into a final design.

Estimating time for such an effort is very difficult. There is no way to force quick results out of creative research efforts. Only the experience of those in control can be used in setting either times or costs, and it is usually safe to make an allowance for this uncertainty. However, as work progresses, the transition from stage to stage will constitute the milestones for administrative purposes.

At the completion of the final design it will be normal to have the corporate management review the whole project before proceeding further. Does the new widget machine appear to fulfill the hopes of Marketing? Should the volume forecasts be changed in the light of intervening market changes? If there are to be changes sought in the product definition or in the specifications, now is the time to make them, because the next step is a major endeavor. The design will next be sent to Production.

In scheduling a major machine development it is difficult to set times for production before a design is available. For the first cut at the schedule, prior experience in production will be the guide for estimating times and costs in the production process. However when the design is available, then Production will look at the design to set its own internal milestones and estimate its costs.

Production will look at the design to determine how it will be made. How much of it will be made in-house and how much will be bought outside? The basic method of manufacture will be selected with due regard to available facilities. Then, for each individual part, the method of fabrication will be selected; will it be cast or machined from solid stock? What will the sequence of operations be? How will assembly be conducted? How will it be tested? How long will it take to procure materials, or new machines if they are necessary? All of these steps are planned, and estimates made of the time to complete each. Cost estimating in production is a little more of a mathematical process, based as it may be on the explosion of the task into many smaller parts for most of which good time and cost precedents should exist.

It must be understood that the foregoing recitation applies to scheduling the manufacture of a new machine. If instead of a widget machine, the project was the production of 500 more typewriters, Model 123, and assuming that Model 123 had been in production previously, the whole development step described above would be omitted, and the production plans would merely have to be reviewed. The planning task would be shortened accordingly. In a job shop environment, where the design is furnished by the party requesting the manufacture, this would be the normal scheduling and budgeting procedure; or if the project was a repeat order for 1000 hub caps of the current model, the scheduling task would be even simpler.

There are two considerations that affect scheduling. The first is called serial dependency, in which the output of one activity becomes the input of the next activity. If a threaded hole is to be produced, the hole must be drilled before the tap is inserted to cut the threads. The time sequence of serially dependent operations must be preserved in any variation of the schedule being considered. The tapping need not be done immediately after the drilling, but it must not be tried before the drilling.

The second consideration is called resource dependency, in which the capacity of the available resource limits the throughput of product per unit of time. From this consideration the time to accomplish the task is derived.

We must not forget that the entry of a new project into the stream of manufacture must be fitted into the available manpower skills, machine loads, material and tooling availability, and all the other commitments of a going factory. The schedule for each project must take its place in the Master Schedule. However the insertion of any new project may adversely alter the timing of other projects. The question then arises as to whether or not the corporate objective and the strategic plans are adversely affected. What it comes down to is that all the projects must be considered simultaneously, together with variations of any one of them, to select the optimum schedule for the totality of the work load.

Note, however, that this process does not at this stage of proceedings go into all the fine details of every step the project will traverse. It merely notes that there will be these steps, and that the overall plan looks rational and feasible. The key milestones of dates and expenditures are set down as targets, and the fine details are filled in, by a similar process, later in time.

Figure 4.1 shows this function in Box 2, labeled "Make Project Schedules and Budgets." It operates under the control of the plans which have been made, and produces the schedule and budget allocations. It also has to adapt to the problems which arise in the course of manufacture, and either deal with them or signal back to management that there will have to be some changes made in the plans. As shown in Figure 4.1, the schedules and budgets are turned over to the Administrative function, Box 3.

## ADMINISTRATION OF THE PROJECTS

The third function of manufacturing management is the administration of the overall operation – the development function, the production function, and the support function. This administration function sees that the plans, schedules, and budgets are carried out, and that the proper resources are available.

It is not enough for the planners to decide what tasks are involved in each project, and to whom they will be assigned. The tasks must be communicated to those who are to do the work. To this end, directives are issued to the ones concerned. Visualize a planner sitting at a desk with a new crop of projects in the in-basket; the planner analyzes each, breaks it down into tasks, notes each task in a directive, and sorts the directives into various out-baskets. Nothing will happen, however, until the contents of the out-baskets are sent to the people who are to do the tasks. This distribution of directives is the control point in the administrative function.

The directives will contain the product definition and specifications, as they came down from the corporate management. They will also include the time schedules for the entire project, so that each facility can see where it fits into the grand scheme of things. The budgeted funds for the project will be spelled out. Most importantly, the directives will authorize the expenditure of time, funds, and resources on the project. This authorization transfers the responsibility to the facility; the ball is in their court, for the moment at least.

The course of true love is never smooth.

Neither is the course of manufacturing. Directives may be modifications of previously issued directives, arising from problems or delays in the work which require a change in plans and hence require a change in the directive.

Administration (Box 3 of Figure 4.1) should receive status reports from Development, Production, and Support (Boxes 2, 3, and 4 of Figure 3.2). The status reports will show the accomplishments to date and the associated costs, so that Administration can compare them to the master plan. The status reports will also show the detailed plans made by each of the facilities for their specific chunk of the task, and any problem areas foreseen. As a result of this forethought and feedback, Administration may adjust the master plan; it may be able to advance the completion date, or it may have to change priorities to expedite matters just to make the planned completion date. When there appears to be an unreconcilable difficulty, it may be necessary to revise the plan, or to refer serious matters to corporate management for their consideration.

"Difficulties," as used here, includes the technological problems as well as schedule or budget problems. It may well be that the performance specifications set up for the new widget machine simply cannot be met, or cannot be met as specified. It is far better to so state, and to get reconsideration and advice, than to persist doggedly on the project and wind up by exhausting the budget and still failing.

An important function of Administration is to provide the necessary facilities for the operation of managerial and administrative personnel all down through the hierarchy. Specifically:

The importance of information processing has been highlighted in this book. This includes means for transferring information over computer networks, storing and retrieving it, and analyzing it for the purpose of making decisions. Hardware and software for data processing, and the training of personnel to use it, must be made available.

Technology is constantly developing, at an exponential rate of growth, and conscious effort must be made to keep abreast of it. The task may be assigned to a person whose duty it is to perceive, understand, and expedite the acquisition of new technology.

In spite of the current attention to automation and to computerization, the human element will always be of paramount importance. Corporate Support (Box 4 of Figure 2.1) will supply assistance, but it is necessary that the environment be such that people are trained and utilized to the best advantage, and organized and managed to bring out their best capabilities.

Product quality assurance is a function that takes place throughout the manufacturing process. QA activities take place in many organizational locations, but Administration must supply the motivation and the support for those charged with the task.

## KEY PERSONNEL

The Manager of Manufacturing has the responsibility of selecting the key staff members of the organization. This includes the heads of Development, Production, and Support, and may include their senior aides as well, depending on the size of the organization. While Corporate Support staff (see Box 4 of Figure 2.1) has trained personnel managers who can handle the procedural matters, and assist in the search and acquisition of the key staff, it is the judgment of the Manager of Manufacturing which ultimately decides on who shall be placed in these key positions. When a key position has to be filled, the appropriate person may be found within the organization as well as outside it; there is a strong preference in most cases to draw top managers from known staff already on the job and familiar with the company and its products.

It is not uncommon to find the manufacturing management function divided into several parallel activities, each caring for the interests of a separate product or product line. The titles "Project Manager" or "Product Manager" are used in some organizations to define a single person with a single mission – to manage a project or a product through its life cycle. Such a project managership comes into being when it is decided by corporate management to enter a new line of activity; it ceases to exist when that line is finally terminated, again by corporate management. For example, a machine tool house may be in the business of making milling machines and machining centers; it decides to enter the business of making lathes. This is different enough to justify a separate project manager for that separate line of machines. Or an automobile manufacturer may elect to treat each year's new model as a separate line of activity, and hence appoint a separate project manager for that model.

The project managers cut across all divisional lines in their concern for their charge, but must operate within the organization's authority structure. Each will want access to all of the facilities of the plant, but must share the facilities equitably and in accordance with the relative priorities of the several projects. The Manager of Manufacturing is the one to settle any conflicts, and to see that each of the project managers operates within the context of the well being of the whole corporation.

## SUMMARY

The manufacturing management function is the first of the four functions which comprise Manufacturing. It wields the driving intelligence and force for the whole manufacturing operation. It makes plans for active projects and products; it makes schedules and budgets; and it monitors and administers those plans.

Each new project must first be understood and then planned. The extent and nature of required research, development and design work is estimated, and the availability of appropriate talent is determined. Times and costs for development are estimated. The nature and extent of required production facilities are estimated and their availability determined for the time periods when they will be needed. If additional machinery or equipment will be necessary, their acquisition is begun.

All of this is summarized in a detailed plan of procedure, and the plans, schedules and budgets are distributed to all concerned. The directives authorize the expenditure of funds, time, and resources on the projects. These directives are the control points of administration.

Each new project is fitted in to the Master Schedule of the active projects, with regard to manpower skills, machine loads, material and tooling availability, and all the other commitments of an on-going business enterprise. At this level, plans do not go into all the fine details which will be generated later by others.

Thereafter, management monitors the performance of the manufacturing activity in relation to the plans. Project managers may be assigned to manage a specific project or product throughout its life cycle. The several project managers must share access to the facilities equitably, and must operate within the authority structure.

The manufacturing manager selects the key personnel for the organization, and supplies them with the necessary tools and skills to manage information resources, human resources, acquire the most up-to-date technology, and assure product quality.

# Chapter 5

# The Development Function

The Development Function is responsible for the development of the product – its origination, perfection, test, and design – and for the maintenance of the design throughout the life of the product. It is solely responsible for the product design, and has sole control over the design.

Corporate management decides what products are to be developed, taking into consideration the corporate objectives, their policies and resources, and the advice of both Manufacturing and of Marketing. Once this high level decision has been made, the mandate is forwarded through Manufacturing management to the Development Function to proceed with the work. This flow is shown in Figure 3.2.

One of the most important elements of the mandate is a good, clear definition of what the desired product is to be, couched in objective, specific terms. If this is not provided, the resultant product may be far from the product that the corporate managers had in mind. Also Development cannot be faulted if there is no clear objective by which to measure their efforts. For example, it is not enough to say "We need a better widget," or "We need a more modern design for our machine," or "We want a machine which will produce a greater rate." How should the widgets be better? More modern compared to what, and by what criteria? How fast, in units per hour?

Another essential is that the product definition should not be so phrased as to inhibit creativity and innovation. For example, if the product definition says, "We need more spindle horsepower in the Model X drill; there is a lot of drill work that machine can't handle," then it is more than likely that the redesign will have a bigger motor and a stronger frame, and little other change. However if the request said "New aerospace metals are so tough that our Model X will not produce holes at rate R," then it may be that different tool edge materials, different cutting fluids, different drill speeds, or some other new metal cutting technology will be the answer, and one that will better meet the user's needs and hence the company's interests.

Development will be evaluated on the extent to which their designs meet the corporate needs and objectives. It is up to the development managers to get a clear product definition of those needs and objectives.

Rarely does any design task start from scratch. An exception to this may have been the lunar landing module in which our astronauts made the first moon landing; there was absolutely no precedent for the design of that vehicle. In general, however, a new product development task starts on familiar ground. It may be only a minor variation on existing designs; it may be modeled on some of the company's existing designs; it may even take as its model the designs of others. Thus to a large extent all development stands upon the shoulders of prior development experience. This collected experience is one of the most precious assets of a company; and because it resides in the minds of the key personnel as well as in the library of the design department, it is one of the most indefinable and intangible of all the company assets. It is also elusive and perishable.

There are two kinds of background knowledge involved. Consider that a manufacturing company is in business to produce a product that will be used by others. The technology of making the product is an internal skill; the technology of users of the product operates in an external, and in a very different kind of environment. Both kinds of experience and technology are essential to a development group. To cite an example from my own experience: The functional design of a piece of shoe machinery involves knowledge of how shoes are made – cut, shaped, and fastened together; knowledge of shoe materials in their almost infinite variety; and of how shoes are fitted and used. The technical design of a shoe machine involves kinematics, cam design, four bar linkage design, stress analysis, force systems, and much more; also, design for manufacturability and an understanding of machine shop practice. Both these two quite different spheres of skill are required in a shoe machinery development organization. The same applies in any development department.

There are other requirements upon the activities of the Development function. The directive to develop a new product is always given within the business context and the constraints of the corporation. Time and funds are not unlimited. Other parts of the corporation must be given consideration, for Development does not operate in seclusion. What is developed must be produced, marketed, and serviced by other parts of the company; and the corporation will usually be found in competition with others, and thus must strive to meet or better their products.

## STRUCTURE OF THE DEVELOPMENT FUNCTION

The inputs to the Development function are the product definitions and specifications, transmitted from top management through Manufacturing management. For new products, the mandate starts a completely new chain of activities, but there may also be requests from Production for modifications in details of the designs to facilitate fabrication, or tooling, or servicing, or materials procurement. Production may also ask Development to design a special tool or fixture and have it made in a tool shop. Along with the product requirements or the change requests will come the authority to expend time and funds on the project.

Except under the simplest circumstances, it is convenient to think of the Development function as comprised of three stages, which may well be reflected in the organization and staffing of the department. Figure 5.1 shows this structure.

The first of the three parts is Conceptual Design, which conducts necessary research and develops the grand scheme for the product, in response to the assigned product requirements. That is shown in Box 1 and labeled "Develop Conceptual Design."

The second part is Preliminary Design, which selects an optimized configuration for the product, refines the concept, reconciles the scheme with reality, tests the design with models, and evaluates its ability to meet the assigned specifications (see Box 2).

The third part develops the detailed designs, including layouts and detailed designs for every part, tests the design with prototypes, and prepares the design for transfer to Production. It is shown in Box 3, labeled Develop Detailed Design.

Normally, a product design does not flow in an orderly manner through the three stages, but loops back to repeat portions of the cycle — sometimes many times — before emerging as the ultimate design. Detailed Design may ask Preliminary Design to reconfigure the prelimi-

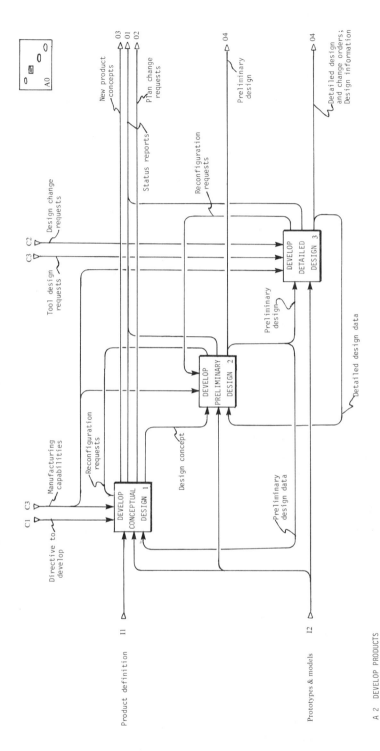

A 2 DEVELOP PRODUCTS

**Figure 5.1.** Develop products (A 2). Decomposition of the "Develop Products": Box (Box 2) of diagram A 0 (Figure 3.2).

52

nary design; Preliminary Design may ask Conceptual Design for a new conceptual arrangement. Requests for detailed part redesign, coming in from Production, will probably go direct to the Detailed Design group. Requests for help in designing special factory equipment, tooling, etc., will probably also go to the Detailed Design group, although there may be a special unit set up for this purpose and resident in the Production facility. Wherever they sit, they are basically design people, and perform a detail design function.

There is a temptation to refer to the three stages of development as Conceptual, Preliminary, and Final Design. However the use of the term "Final" applies only in the sense of the sequence of the three stages. As we shall see, a design is rarely final when it leaves the third stage. The more definitive term "Detailed Design" has been used to describe what happens at that stage.

An important control on Development is the limitation of the equipment, facilities, and skills available to the Production function. Information on this subject is fed back to Development from Production, and made available throughout the function.

Let us now look in detail at these three developmental functions.

## THE CONCEPTUAL DESIGN STAGE

The first of the three stages of the developmental function is the conceptual design stage. It is the beginning of the life cycle of activities included in the meaning of the word manufacturing as used in this book, as explained in Chapter 1. It is shown as Box 1 in Figure 5.1, named "Develop Conceptual Design." It conducts research into the needs of the users of the proposed product, and any applicable technology. It then selects a tentative product configuration to meet the needs of the users and the specifications of the directive to develop the product.

This function is more closely focused on the technology of the company's product than on the technology of producing that product. For example, if a company is making spinning frames for the textile industry, its research and conceptual design might be most concerned with new textile fibers and their behavior, and on optimum doffing cycles to achieve maximum throughput with minimum operator attention, and similar problems. The problems of designing very high speed bobbin spindles would be left to later stages of the development function. If a company produces candy making machinery, it may focus its research on control of chocolate mixture consistency in the enrobing machine, and leave the design of thermostatic controls to the detail designers.

The conceptual design function first identifies the need for information not already available in the company, and seeks sources therefor. The question is, "Do we have available technology to meet the need that has not already been applied in our product design?" This calls for a search through the research department files and the staff's minds. It also calls for a literature search outside the company, to find any pertinent facts which are publicly available. Patent Office files may be searched for information. Usually a company research facility has a person assigned to do such searches, and in larger companies this may well be a full time job for someone.

It is expected that workers in a research laboratory will, as part of their professional responsibility, keep abreast of their field through participation in professional society meetings, seminars, and committee work. This helps to acquire knowledge of what others in the field are doing, and sometimes long before the material ever appears in print for general distribution. Of course, this "corridor gossip" works two ways, and care must be taken to protect advanced in-house information which if disclosed would have an adverse effect on the company's competitive position. Large companies may have a routine surveillance by a patent attorney of information to be released.

Pertinent research may also be conducted in academic institutions or research centers maintained by trade associations or similar groups. This is generally freely available but must be sought out. Large companies may sponsor research projects in colleges or universities, funding the student/professor team and their laboratory expenses. It is customary for the sponsoring organization to designate the subject area and general direction, but the academic institution will insist on its freedom to direct the work and to report thereon.

It is to be presumed that the staff of the research laboratory will keep themselves informed of all available information in their field of technology, and stand prepared to apply it to any assigned development project. Trips to observe the company's products in action offer an excellent insight into real world working conditions, and a better understanding of what the company products must do. The Marketing function can make the appropriate arrangements.

All of this collected technological information is compared to the assigned developmental task. If the task can be accomplished without additional information, the conceptual design function moves to the next step. However if there is an additional need, a research project in-house will then be started. Laboratory models may be requested for the purpose; Detail Design can make up adequate sketches, and Production can put them through the shop and deliver the models to the researchers. As experiments progress toward the needed extra information, careful records are kept to establish conception dates for later patent action.

Out of all this will come a concept for a new product that appears to meet the product definition. (If not, then the review by higher authority is in order, either to modify the definition or to abort the project.) Then the concept will be tested for feasibility, insofar as model tests will establish that fact. It is not unlikely that alternate concepts will have developed in the course of the work, and they will be compared to determine their relative attractiveness. Sometimes two equally attractive alternatives will be sent on to Preliminary Design to select the preferred one.

Also sometimes, out of all this, a concept emerges for a new and different product quite outside the bounds of the assigned product definition – an inspiration such as the Patent Office refers to as a "stroke of genius," but which would more properly be labeled an astute perception. Such bonus ideas are documented, and then referred back to corporate management to consider for separate authorization.

The mechanism for transfering product concepts to the next stage of development – preliminary design – is hard to define. Models and demonstrations, sketches, descriptive documents and consultation all play a part in the process. Not infrequently the chief worker may move with the project into the preliminary design stage, carrying with him all the elusive bits of knowledge which would be hard to write down, particularly those ideas which were considered and discarded for good reason. He will also have a strong motivation to see his concept succeed.

Finally, the preliminary design which evolves will be run back past the conceptual design workers for their information and for their further guidance.

## THE PRELIMINARY DESIGN STAGE

The second of the three stages of the developmental function is the preliminary design stage. It is shown as Box 2 of Figure 5.1, labeled "Develop Preliminary Design." It focuses on the selected concept, and by modeling, testing, and analysis, refines the concept into a design ready for final detailing. It builds confidence in the ability of selected configuration to meet both the technical and the economic requirements of the project. When there are alternate concepts submitted, it chooses the one most likely to meet the product definition. Before committing the preliminary design to the next stage of the process, the preliminary design will pass critical examination to assure all concerned that it is indeed a suitable base on which to continue.

If the product involved is uncomplicated, for example, the addition of another size of bolts to an already existing line of sizes, this and the preceding stage would be fairly brief and straightforward. However if the product is, for example, a complete machine of a new sort or a product of similar complexity, then the preliminary design stage may be lengthy. We will examine the preliminary design stage on the assumption that the latter case is involved, so that we may explore all the ramifications.

Formulating a preliminary design is one variety of problem solving. The desired end is prescribed. The available knowledge is at hand. The task is to find the bridge between them. The procedure is logical: You attack the problem by all known analytical methods; you subdivide the problem into subproblems and sub-subproblems, and analyze each. The process is like peeling an onion, removing one layer after another until what is left is just one single little core. That little remaining nub of a problem may not yield to analysis or any other of the usual thought processes. An invention is needed.

Many years ago a Yale professor named Eliot Dunlap Smith discussed his theory of problem solving. He used as an analogy a kaleidoscope — a child's toy which generates an infinite variety of colored patterns for visual examination. It is made from a prismatic array of three mirrors; the eyepiece is at one end and a glass closes the other. Behind the glass is a narrow space before another glass barrier. In the space is a handful of varied glass beads, bits of shell, and other small objects. When the glass end is held up to the light, and the user looks through the eyepiece, the randomly arranged bits of translucent objects are reflected in hexagonal symmetry to form beautiful patterns. Shake or turn the toy, and a completely new pattern appears. No two patterns are ever alike.

Professor Smith likened the kaleidoscope contents to the mind of the problem solver — a lot of bits of information filed away waiting for use. The process of solving that ultimate nub of a problem is one of trying one pattern of possibilities after another, discarding one after another, or changing the patterns a little to improve their acceptability, just the way the kaleidoscope user shakes down pattern after pattern until he gets one he likes. Smith's point was that solving the ultimate problem is not one that can be solved by forcing an analysis; it takes a flash of perception. The mind sorts through all its background store of knowledge, looking for the right combination. One can not force this process; in fact, it helps to get the problem clearly stated and then turn one's attention to something else – golf or gardening or some other thought pattern. If the answer is going to be found, the subconscious mind will keep working at it; if an answer is within your capability, it will sooner or later bob to the surface of consciousness. If the answer does not lie within your powers, nothing will happen.

Smith's second point was that if there were no bits of glass beads in the kaleidoscope there would be no patterns, no matter how hard the child shook the toy. Also, if an engineer has no store of bits of background information, no matter how hard he tries to set the stage, nor how long he waits for perception to strike, no answer will appear.

The preliminary design stage begins by expanding the conceptual design into a detailed structure of subsystems and sub-subsystems. Together they make a complete machine, even though the concept received from Research may have set forth only the novel elements of the new product. Every machine will need a frame and a power supply and a control, things which the concept design probably had simply assumed would be available. The technical and economic suitability of each subsystem is proven, by test if necessary, and by analysis. Prototypes and models may be built for this purpose; they will be good enough to perform at full speed and load, at least for a short time. To get these prototypes, Preliminary Design will have drawings made and sent through the model shop.

The relationship of each subsystem to its parent system, and of all the various subsystems and sub-subsystems to one another is tested to see that they will ultimately coalesce into a reliable working machine. Then the total machine is tested and evaluated for conformance with the product definition and specifications, to see that indeed the assigned objective has been met. The machine will be tested over the full range of its possible conditions of use, to make sure that when it gets into the field no one will wander in with some application which it ought to be capable of handling, but for which it was never tested.

The final prototype may well be tested in the field, under real world operating conditions. It always amazes designers to see how brutally field conditions may test a product, probably because it is beyond the comprehension of a machine designer that anyone would treat a machine the way some operators do. Field tests will require the cooperation of Marketing, because contact with potential buyers is involved, and arrangements must be made diplomatically. The patent counsel will also have to give his approval for field test of a product, because that constitutes "public usage" and the time clock starts running down the period within which any patent application must be filed.

It is almost certain, except in the simplest of cases, that the preliminary design will go through a number of iterations of the foregoing process. There will be disappointments and failures to be corrected. It will become apparent that there is a better configuration than the original best guess at configuration. Tests will be miserable failures, and changes will be required. The whole preliminary design process, or parts of it, may be reiterated many times.

Development managers will keep one eye on the budget and one on the test results, because funds are never inexhaustible. There is a strong temptation to think that just one more change and one more trial will achieve success, and it takes real courage to stop and go back to Development management for either a change in the whole machine concept, or a critical review to see if perhaps the objective set up is unattainable within the time and budget available, and should be changed, postponed, or aborted.

An important consideration in the preliminary design stage is the producibility of the design in the company's facilities and with their available technology. It is obviously undesirable to predicate a new machine design on a production method which the company does not have. If that cannot be avoided, then the Production people will need some advance warning to get prepared. The cost of new facilities must be weighed in the decision to accept such a production method, as against some less elegant method which is already available. It is increasingly the practice to have someone from Production closely allied with Preliminary Design just to make sure that in-house producibility is considered. Such a liaison person may be detailed full time to Development from Production.

This close feedback from Production to Design is particularly important in fields where production technology is developing almost as fast as design technology. The electronics industry is a good example. Newly developed production abilities may justify redesign of a product while it is still in active production, simply because the new technology will make possible a new capability in the product itself.

As shown in Figure 5.1, there are two feedbacks from Develop Detailed Design, Box 3, to Preliminary Design, Box 2. Detailed design of the machine parts may uncover a place where the configuration would be easier to make or better if it were changed a little, or where Production or Support have asked for changes in the detailed design. Second, Preliminary Design is kept informed in general on the work of the detail designers; this information constitutes an input to the further work of Preliminary Design.

Development management will authorize the transfer of a project from the preliminary design stage to the final or detail design stage.

## THE DETAILED DESIGN STAGE

The detailed design stage of the development function converts the preliminary design into a detailed and finished form suitable for use in manufacturing. When the preliminary design has demonstrated that it meets the product definition and the specifications, and appears to be producible in the factory, transition to detail designing is authorized. This final stage of the development process represents a not inconsiderable expenditure of time and effort, and the step is not taken lightly.

Detailed Design will receive from Preliminary Design the layout drawings, such part detail drawings as exist, the models and the test results, and access to the people who developed the preliminary design. Because development is an evolutionary process, it will turn out that some of the layouts are either missing or out of date, that the detail drawings may have been changed several times in the course of the experimental work, and that a few more changes suggested by the final trial of the model have yet to be recorded. It is the task of Detailed Design to take this bag of rough nuggets and refine it into a final design. It will be final, that is, until experience in the factory or the field dictates some further refinements. These will come in the form of design change requests, as shown in Figure 5.1.

The detailed design must be so complete a definition of the design that other people, at other places, and at other times, can take the drawings, layouts, and bill of materials, and without any additional instruction or advice, fabricate and assemble parts to make the intended end product – a machine that performs its intended function reliably and creditably in a real-world environment. That is no small order.

The first step in this stage of development is familiarization with the preliminary design and its records. It is expected that there will be access to those who have worked on the project in its earlier stages. Sometimes the same crew continues on into the detail design stage, merely switching their hats around to designate the transition, in which case the transition is easier.

The second step is to make a set of final layout drawings (or their equivalent if the product is not a machine), to show the overall configuration. There will be layouts for each machine subsystem, and for each sub-subsystem, to as many levels as is necessary to make every bit of the machine's configuration completely clear.

Layouts, too, are products of an evolutionary process. Layouts show an assembly of parts, and the parts to be drawn must have shape, dimension, and mass. This raises two questions: Are the parts as drawn strong enough and rigid enough to perform their function? And if parts of the mechanism move, do they have proper clearance from other parts at all times? Here we have a "hen and egg" problem. If a part turns out not to be strong enough, the answer may be to make it bigger; but if it is enlarged, will it fit into the space available? What this comes down to is that the designers go through a sequence of design calculations on every important part, adjusting dimensions, materials, and fabrication techniques until each part is adjudged to be capable of meeting and performing its function; and further that each part will have room in the mechanism to move if it is supposed to, and in any case will mate with the adjoining parts.

The advent of the computer has revolutionized this phase of design. Strength members have to be analyzed and sized for static strength, for stiffness under load in order to resist deflection, for natural frequency of vibration, and such like characteristics. For simple shapes like beams, cylinders, and shafts this is a task which can be solved with the formulas of applied mechanics and a table of logarithms. However, for complex shapes, the mathematics, even if understood, is too complex and the calculations too laborious to be practicable. Lacking the power of the computer, the designer of old made his best guess, added a "factor of safety" (which meant multiply by 10 or more) and let the design go to test. A trial part might be made and tested to destruction to prove its strength. If the part broke in use, the answer was very simple: make it stronger, and try again.

For kinematic mechanisms the procedure was similar. Force, torque, and power calculations had to be made. It was necessary to calculate the weight, center of gravity, and moment of inertia of the parts, and the force reactions on one another and their supporting parts. Trains of parts transmit motion and force from the prime mover to the working point, and impose loads on the stationary members of the mechanism, which may in turn require the recalculation of those members.

All these laborious and sometimes arcane calculations can now be done with the aid of computers, many times faster than by hand; and many computations too lengthy or too laborious ever to be done by hand are now ripped off in minutes by the digital computers. For example, the main spar of an aircraft wing is a very complicated piece of design work, but by using a routine called finite element analysis, the stresses and deflections can be calculated to a high degree of accuracy. The spar of a well known aircraft was designed in this manner. The analysis involved over 200 simultaneous differential equations, and the data base contained over 2.5 billion bits of data! The wing spar is as strong and as stiff as it needs to be, and as light as possible consistent with safety. Such calculations were utterly impossible before the advent of the computer.

The next step in the detailed design phase is the creation of detailed part drawings for every part. Trade jargon calls this "detailing." Some parts such as standard hardware – bolts, nuts, washers, pins, some bearings, and the like will be shown on the layout but will not be detailed, because standard drawings for them are to be found in vendor's catalogs and in the company's drawing room files. Purchased items such as ball bearings, motors, hydraulic fittings, push buttons for control panels, and the like will also be specified but need not be detailed; the vendors will supply drawings if drawings are necessary.

A detail drawing is a piece of paper on which is recorded all that is necessary to completely define the part to which it applies. There was a custom years ago to use large sheets of paper for details, and put drawings for as many parts on that sheet as could be fitted in. Happily that custom has been abandoned; only one part is shown on one drawing. The drawing has a number, which also becomes the part's number. If, in the course of time, the part is revised or withdrawn, the part number stays with it. The detail will have a title box containing the part's name and number, together with the company's logotype.

Figure 5.2 is a simple detail part drawing. In addition to the name and number, the title box contains the date of its creation, the dates of revisions, the draftsman's name, and his supervisor's name or initials. It shows the scale of the drawing. It is desirable to show the part in full size, but many parts are so large that they could not be shown on a manageable piece of paper, so they are drawn in reduced size; the reduction is the "scale" of the drawing.

The part is shown on the detail in a conventionalized pictorial form. Parts are three-dimensional objects, and the drawing paper is two dimensional, so an orthogonal projection of the part is drawn, showing front, top, and end views, plus any cross sections or other views as are necessary. People who use part drawings can recreate in their mind's eye the solid image from the paper drawing.

The size of the part is defined by dimensions added to the geometric pictorial. Witness lines extend from the points on the drawing between which the dimension applies, and the dimension lines run from one witness line to the other, with the numerical dimension written in that line. Now it is important to note that in the real world, it may be difficult to make a part with the precise dimension specified; tools, materials, and workmen vary, and parts supposedly identical will, after manufacture, turn out to be slightly different from one another. Some variation is tolerable, and is anticipated in the design. Modern machine tools can work to within 0.005 in. either way from the intended dimension, but when closer tolerances are required, they may be so specified. The toler-

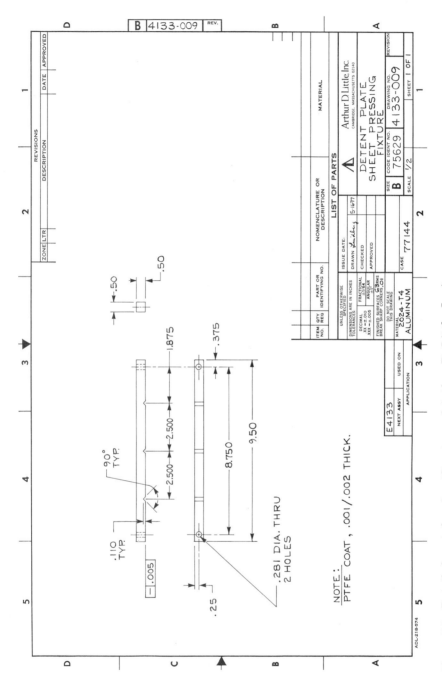

**Figure 5.2.** Detailed part drawing (Courtesy of Arthur D. Little, Inc.).

62

ance will be noted on each dimension; if no tolerance is noted, there is a default tolerance indicated in the title block.

When the factory comes to make the part, it will not use its imagination; every surface and every location depicted on the drawing must be specified as to its position relative to all the other features. To fail to do this will leave some one in doubt as to just where to cut or form some surface; this will not do.

In addition to geometrical configuration of the part, shown by the pictorial, and the size of the elements, shown by the dimensions, it is necessary to define the material for the part. Whether it is made of metal, wood, plastic, or some other material, the material is specified by reference to standard materials codes, or names, and for unusual materials the source may also be noted. Metals may be hardened or annealed, so the hardness is specified either in a hardness code number or by implication in the material's own code number.

Materials may come in a variety of surface finishes, and if any part of the raw material is to appear untouched in the finished part, that surface texture will be specified. The surfaces left on the finished part by the fabrication operations will also be noted on the drawing. After fabrication to the correct shape, the part may have to be hardened or annealed; the surface may have to be cleaned, polished, plated, or painted. All these instructions are printed on the drawing.

Parts which are to be purchased already finished from an outside vendor may require some minor shop operation such as numbering, painting, or some such operation. Drawings may be a convenient way to record this information.

All of the above long litany about what goes on a detail drawing is offered to reinforce the concept of the importance of these documents. They are the only part of the design except the bill of materials which will ever get into the shop, so they must be complete and unambiguous. And needless to say they are expensive to make. Manual drafting methods will produce, for a very rough average, a detail in 4 to 40 hr, depending on its size and complexity; and there will be a drawing for every part that is to be made.

Is there a way to reduce the cost of detailing? There are two. One obvious way is never to draw the same part twice. This applies to a company making a variety of similar products, or a succession of similar models of the same product. A little study will show that there are some parts which are designed over and over again, with only minor changes, if any. A large aircraft design department had a display board with sixty parts on it, enough alike to be confusing. Each was a bracket for attaching something to a bulkhead in an airplane. Each was made by making a right-angle bend in the middle of a rectangular sheet of metal. Each had two or three holes for bolting it to the bulkhead, and two or more holes for attaching whatever was to sit on it. The widest range of dimensional variation in all the parts was no more than two to one. It was estimated that each part separately had cost over $2000 to draw, engineer, tool, and prepare for production. One "family of parts" drawing would have done for the whole lot, and any one of the family could have been specified by tabulating a half dozen dimensions on a typed sheet. Figure 5.3 illustrates such a drawing.

The technology that makes possible the avoidance of repeated design of the same part is known as "Group Technology" or as "Classification and Coding," a better description. There are several commercially available systems which teach this technology. In essence each is as follows. Parts are classified in accordance with a number of criteria, by a sequence of numbers. For example, parts may be turned parts, box like parts, sheet metal parts, etc. Turned parts are classified as either uniform in diameter, larger at one end than the other, or larger in the middle than at either end. Next they are classified as to whether they are solid or hollow, and if hollow, whether the hole goes clear through. They are next classified as to length, sorting them into one of several

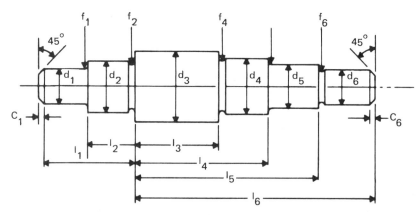

**Figure 5.3.** Family-of-parts design drawing. This drawing is applicable to any shaft with two steps to the left, and three steps to the right of the central major cylinder, or to any shaft with a lesser number of steps on either side. $f_1$, $f_2$, . . . are codes for the radius of fillets and root diameters relative to the smaller adjacent step. Individual part drawings are not drawings, but tabular entries of dimensions as coded on this master family-of-parts drawing.

length spans; and again by diameters, and by any other significant parameters.

Each classification is given a code number, so that if there are eight criteria of classification, an eight-digit number will describe a highly unique part configuration. This is the coding part of the technology. A company classifies and codes all the parts it has ever designed. Classification can be detailed enough so that only a dozen parts are found in any one class. When a new part is to be detailed, the part is classified and coded, and then reference is made to all the previously designed parts of that code number. The detailer calls up on a computer printout or on a CRT screen the old parts, one by one, to see if one of them may not be what he has in mind, or close enough so that he can adapt to its use. This will save an entire detailing job. If he does not find exactly what he wants already in the files, the chances are excellent that he will find one very close to it. He can call up that part design, and by modifying only one or two dimensions, he will have his task done.

This same data base of classification and coding will be useful many times in the subsequent stages of development and production. The technology is a powerful tool for economic design, just now (1980s) being appreciated and adopted.

The second way of reducing the cost of detailing is to switch from drawing parts with pencil on paper, on a drawing board with T-square and triangle, to the use of interactive computer terminals. The detailer is equipped with a computer terminal having a keyboard and a CRT screen. He also has a mechanism for describing the point on his "drawing" where a line is to begin or end. He "draws" the detail on the face of the tube, where every line and curve appears at the point he designates. Lettering is easily added, as are dimensions. This has come to be called CAD, standing for "Computer Aided Design." Figure 5.4 shows such a piece of equipment. When the drawing is complete, the detailer can transfer the data set which he has created to a plotter, which will draw out the paper version of his design. The design may be retained in the computer's memory, or transferred to another computer for further processing.

A number of vendors offer CAD equipment, some of great versatility and complexity. Users claim from three-to-one increases to as much as eight-to-one increases in the output per man, using such terminals. The art is in its relative infancy, and offers great opportunities for closer integration of the development and the production functions.

The next step in the detailed design function is to check the previous work by creating assembly layouts, starting with the detail drawings and assembling the parts into the total machine configuration. Any er-

**Figure 5.4.** Interactive graphic terminal for computer aided design work. The Designer V System, courtesy of the Computervision Corporation, with its CRT screen and controls in the center foreground. A cylinder printer for plotting designs on rolls of paper is to the left, and a large drawing board/digitizer for taking input from an existing drawing is at the right.

rors in dimensioning the parts will show up on the assembly layout as an interference, and will call for corrective action. A check is made for tolerance accumulation, meaning that when all the individual part's tolerances are at their collective worst case position, will the machine still go together and work?

Assembly drawings also may act as guides to the factory in the actual act of assembly, and will assist the Support Function when they are preparing operation and maintenance manuals. "Exploded" versions of the assembly layouts, unit by unit, show the relative position of the parts, a great help to users and service men in identifying parts for service or replacement. Such an exploded view drawing is shown in Figure 13.1.

Finally, the detail design stage prepares one of the most widely used documents in the entire manufacturing activity – the Bill of Materials.

It is sometimes called a parts list, or an indented parts list. The bill of materials is an hierarchical listing of everything that goes into the final product. Its structure follows the hierarchical structure of the overall layout, the next level sublayouts, and so on. Figure 5.5 is a sample taken from a bill of materials.

There is a line in the listing for every part, giving the part's name and number, how many parts are required, and similar information. The left end of each line is indented a few spaces to show the hierarchical relationship of the assembly pattern. A major part has a flush left margin; its major components are indented one space; their components are indented two spaces, and their components are indented three spaces, and so on. A group of consecutive parts with the same indentation are to be assembled into the part on the next line above having one less indentation. Fasteners and purchased parts as well as designed parts are listed.

Bills of materials used to be hand lettered on printed forms, making any revision a laborious (and obnoxious) task. The advent of the computer word processor technology has made the task very much easier and therefore very much more amenable to revision and updating.

| NAME | | | | | | NUMBER REQ'D | | PART |
|---|---|---|---|---|---|---|---|---|
| 1 | 2 | 3 | 4 | 5 | 6 | NEXT HIGHER | TOTAL | NUMBER |
| | Eccentric Gear | | | | | 1 | 1 | SA 924 |
| | | Gear balance weight | | | | 1 | 1 | SA 921 |
| | | | Eccentric pin | | | 1 | 1 | SP 922 |
| | | | | Pin lock nut | | 2 | 2 | NL 4711 |
| | | | | Rivets | | 2 | 2 | PL 6730 |
| | Slider Assembly | | | | | 1 | 1 | SA 959 |
| | | Slider block | | | | 1 | 1 | S 9590 |
| | | Counterweight track | | | | 1 | 1 | P 6152 |
| | | | Counterweight track screw | | | 4 | 4 | SL 9396 |
| | | Counterweight | | | | 1 | 1 | S 5951 |
| | Plunger | | | | | 1 | 1 | S 9591 |
| | | Slide block track | | | | 1 | 1 | SA 928 |
| | | Blade clamp | | | | 1 | 1 | SA 2281 |
| | | | Blade clamp screw | | | 1 | 1 | S 9346 |
| | | Blade | | | | To order (See catalog) | | |
| | Plunger Housing | | | | | 1 | 1 | A 118-3 |
| | | Plunger housing bushing | | | | 2 | 2 | S 9589 |
| | | | Plunger bushing screw | | | 2 | 4 | SL 922 |
| | | Plunger housing felt | | | | 2 | 2 | F 6862 |
| | | Plate | | | | 1 | 1 | P 7072 |

**Figure 5.5.** Typical bill of materials. Extract from a bill of materials for a saber saw. Part names are printed in the indented format to aid the user in visualizing the hierarchical structure of the list.

It will also do aggregation, meaning that if say, ½-in. cap screws are used in a dozen or more places, each time calling for the number needed at that particular spot, the computer will look through the whole bill of materials and aggregate the number of ½-in. cap screws needed for the whole machine.

There is probably no more important nor widely used document in manufacturing than the bill of materials. When Marketing offers a manufactured product to a potential buyer, a bill of materials will be used to define the principal components of the machine; on this item by item listing will be based the price of the machine. When the project is committed to preliminary design, the bill of materials is greatly expanded to show the configuration of the machine. When the detailed design is released to the factory, a working bill of materials will go with it. Manufacturing engineering will base its manufacturing plan on the major components of the bill, and its part production planning on the line items in the bill. Assembly will be guided in planning its procedures on the hierarchical structure. Purchasing will order parts and raw materials on aggregates taken from the bill. A bill of materials will go with the machine when it is shipped, for guidance in operating and maintenance.

Each of these many users of the bill of materials will look for something different in the document. This is where the versatility of the computer documentation pays off. The computer will extract and organize the data in any way convenient to the specific user. Furthermore, when changes are made in a machine during its life cycle, the bill of materials is corrected to show the changes, and any subsequent person needing information will, when he gains access to the record, get up-to-date information.

In the course of detailing it may well transpire that the design layout contains parts which could be improved in producibility or serviceability if they could be changed. In this case the Detail Design function will ask the Preliminary Design function for permission to make a modification. Similarly, when a set of details gets to Production, the manufacturing engineers may ask Detailed Design for a change in some parts. Again when the machine gets into use, service may report problems in low reliability, difficult disassembly and replacement in the field, or suggest some functional improvement. All these change requests come back to Development for action. When only the detail design has to be changed, the Detail Design group attends to the matter. However if major changes seem necessary, higher level consideration and approval is necessary.

When a design change is proposed, a decision must be made as to the expected benefits versus the cost of making the change and of changing all the downstream records, tooling, and stock in process and in inventory. If it is decided to make the change, the timing of its introduction is also decided; it may be made mandatory immediately, or to take effect as current work in process or stock in inventory is exhausted. In addition to the cost of changes, potential product liability problems must be considered.

It must be remembered that Development has the sole responsibility for the design of the products; it must therefore have the sole authority to change the designs. It keeps track of all revisions so that it can identify the revision level of any part in any machine in the field, so that if a call comes for a replacement part, the part sent will be identical to the one to be replaced. Again, the data handling capabilities of the computer serve well in this regard.

## QUALITY ASSURANCE

There is little profit in producing defective products. If a product's defects are apparent, it cannot be sold; if its defects are not obvious, the buyer will be seeking redress as soon as the defect is perceived. Repetition of such off-quality production will damage a company's reputation far beyond the value of a single unit of production. It is, therefore, most important to assure that all the products of a company are of good quality.

Both quality and productivity are highly desirable, but quality production requires an effort, so it is easier to maintain productivity if quality is not a concern. There is, therefore, always a compromise to be achieved between high quality and high production. Quality is most economically achieved if the product is so designed that it can readily be made correctly, and the production methods are selected to assure that it be "made right in the first place." On the other hand, quality may be achieved by carefully inspecting the finished stream of products and culling out all units that do not come up to the quality standards. The first of these two procedures is called quality assurance (QA) and the second is called quality control (QC). Both are essential, but the better the QA, the less the cost of QC.

In practical operations, the inspectors who exercise the QC function sit as judges over the work of their fellow operators on the production line. To prevent interpersonal friendships or enmities from prejudicing their judgement, or to prevent a senior official in the production department from overriding the rejections of an inspector in the interest of meeting a shipping deadline, it has been common practice organizationally to have the inspection department report to someone senior to the person in charge of production. This has tended to separate the QC department from the rest of a plant hierarchy.

The end result of this structure has been to set the quality control function up as a separate organizational entity. This may indeed be necessary, but the quality assurance is not a separate function. It is an intrinsic component of every part and every function of the manufacturing operation. It has, therefore, not been shown as a separate box on the functional diagrams; it is more properly treated as an obligation on all functions.

Conceptual design will select product configurations that have the greatest inherent reliability in use, and adopt technologies for the company products that are known to be sound. Preliminary design will select mechanisms that are known to have high reliability and to be producible on the facilities available. For example, if precise location of a rotating part is essential, ball or roller bearings may be selected in preference to journal bearings. As another example, in an NC machining center it may be required that the work table be positioned with extreme accuracy – say 0.0001 in. – and that experience has shown that in ordinary production this is difficult to obtain, and even harder to maintain during prolonged use. A self-compensating mechanism may be designed into the table control system which, when calibrated with a laser interferometer, will correct the local error at each point of the motion.

Detailed design will take care that the detailed drawings fully describe the wanted parts; that tolerances are tight enough so that the assembled mechanism will function as designed, but no tighter than necessary in order to avoid cost.

Machine tools will be purchased with an eye to their ability to maintain dimensions in part after part of mass production orders. Machines may be selected with self inspection capabilities that watch the drift of critical dimensions from part to part, and correct the tool setting before the dimension gets out of tolerance. Tool maintenance will be carefully and regularly done, to make sure that the wear and aging have not detracted from the tool's performance. Operators will be trained in the proper use of the machinery, in order that it be used to the best advantage. They will also be encouraged to inspect their own work as they proceed. Incoming materials and components will be carefully inspected to make sure that they conform to the purchase order specifications.

All of these actions are means to assure quality being built into the product: "Make it right the first time." However there is one additional task that must be done to be completely sure of turning out quality products. This is quality control. There is a saying in the industry that

"You can't inspect quality into a product." However one can ensure that poor quality does not get out. The actual production must be inspected to catch the inevitable errors that creep in, in the best of environments. Inspection operations are just as much a part of the production sequence as is the shaping of a part. The earlier in the production sequence they are caught, the less the cost of either correcting the error or scrapping an unsalvageable part. So the design of the inspection operations itself is a quality assurance function.

## THE OUTPUT OF THE DEVELOPMENT FUNCTION

Figure 3.2 shows the A 0 diagram of our functional structure of manufacturing. We have been discussing the Development Function, shown in Box 2 of that diagram. The major output of the Development Function is there shown as the transfer to "Produce Products," Box 3, of the product design (and of any changes in the design that may have occurred).

The designs act as a control for the Production unit, rather than as an input, because they are not consumed and passed along, but remain after the true input to Production, the raw materials and other procurable items, have been converted into products and shipped out.

This points up an interesting distinction between data (the control) and materials (the input). Data are infinitely divisible. If person A has a piece of data, and tells it to person B, A still has it as well as does B. A can go on, telling it to C and D and more. Then they all have all of it. However material is different. If A has 100 bolts, and gives B 50 of them, then A has only 50 left, and B has his 50. If A gave B 100 bolts, he would have none and B would have all. People who are accustomed to the numerical accountability of money and material objects have to adjust their thinking when dealing with data, and designs are basically data.

There is another interesting feature about data as compared to material. Partial transfer of data is no transfer at all. If a person says "Meet me on arrival at O'Hare at 10:15 AM," it will be meaningless unless the date of arrival is also given. However, as in the example above, when A gave half the material objects (bolts) to B, they were in B's possession and useful to him. If a detail drawing is made with missing dimensions, it will be no better than no drawing. Data transfer must be of complete data.

Development makes information on product designs available to all who have a need and right to know. This includes the Support function (Box 4 in Figure 3.2), as well as Production. The direct recipients in Production will be those who do the planning for manufacturing and the planning for production, Boxes 1 and 3 of Figure 6.5. Both of these recipients may also be informed of the design when it emerges from Preliminary Design. There are many instances when a little advance notice will permit those who order in materials to place orders for very long lead time items. For example, in designing a new airplane, orders are tentatively placed with jet engine manufacturers at the same time that design commences on the rest of the plane, so that the engineers may be ready reasonably near the time the plane is built. There are also many advantages to be gained by feedback on the manufacturing facilities available in the factory, and criteria for producibility which the manufacturing engineers can offer.

Transfer of a design from Development to Production requires the action of Manufacturing Management, and sometimes of Corporate Management. It commits the company to an expense for preparation for manufacture which may be as large as the expense already invested in development. It requires a review of the company policies and programs, the current status of the market, and the chances of profiting from the manufacture and distribution of the new product. In short, this step is a major Go/No decision point.

## SUMMARY

The Product Development function proceeds in three stages:

1.  A product concept is evolved by application of known technology to the product definition, conducting research if necessary to find the necessary technology. The concept is embodied in a configuration of the product.
2.  The product concept configuration is converted into a preliminary design, parts or all of which may be modeled and tested to prove that the design meets the specifications and appears to be producible.
3.  The preliminary design is converted into a detailed design consisting of layouts, part details, and a bill of materials which collectively contain all the information necessary for the production of the parts and the assembly of the product.

# Chapter 6

# The Production Function

The Production Function converts procured raw materials into finished products in accordance with designs received from the Development Function, and delivers them to the Marketing Function, all in response to directives from the Management Function. This relationship is shown in Figures 3.2 and 2.1.

Let us carefully distinguish between this function and the organizational unit and its structure which frequently bears this same name and performs this task. The organizational unit is frequently referred to as "manufacturing," "the manufacturing division," or "the factory." In this book we have agreed upon a different scope for the term "manufacturing," see Chapter 1.

The "factory" is usually sharply differentiated from the other parts of the corporation – Research and Development, Engineering, or Sales – by a number of criteria. It usually has a separate organizational structure, leading right up to a Vice President. That officer ranks with the Vice Presidents of Sales and of R & D, the Treasurer, and similar officers. Manufacturing is frequently geographically separated from the other parts of the company, particularly when the type of activity dictates a location in an industrial area of the community, or adjacent to a waterway or a railroad, or close to the source of raw materials.

The manufacturing unit usually has a large number of employees relative to the other parts of the company, and hence is located near an appropriate source of labor. When the company is a multiplant organization with several factories, this geographic scattering is even more pronounced. (The functional point of view regards all the factories in a multiplant organization as all performing the same production function.) Large employment rosters mean many levels in the management hierarchy, and hence many middle management people.

Finally, the manufacturing unit uses the largest fraction of the company's capital assets. Manufacturing is capital intensive; plant buildings and manufacturing machinery are expensive relative to offices and office machinery. Added to the investment of facilities and machinery are the inventories – raw materials, work in process, finished goods, and tooling.

All of these differences, taken together, tend to set the "factory" apart from the rest of the organization. While it is a good thing to distinguish the several functions, it is not desirable to set one organizational unit apart from all the others by any criteria, since it leads to the creation of real or psychological barriers between them.

There is, however, one very real distinction: The Production Function is the only unit which is directly involved in the physical handling and converting of raw materials into finished products. The material conversion activity seems to be the dominant concept connected with "manufacturing" in the popular mind. As this book demonstrates, material conversion is only one part of a much larger whole.

The functional structure of Production is not the same as the hierarchical structure of the factory, nor should it be. The hierarchical structure, as epitomized by the usual organization chart, is a good tool only for the control of people. There was a time when people had the control of technology; any and every step in manufacture was the real-time output of human intellect. Ideas in the mind of a person were drawn out on a piece of paper by a person. Plans for production were made by people; material was handled by people and the machinery of production was controlled by the mind and hand of a person, step by step. When this was true (and it still is to some extent in some places), the organization of the manpower was in effect an organization of the production function. Today, and increasingly, this does not hold true. Control must be exercised over not only people, but over machinery, materials, methods, and money. An organization chart will not adequately determine the behavior of any of these but the people.

The Quality Control function is a good example of the difference between a company's organization and its functional structure. Because errors inevitably creep into manufacturing, it is necessary to inspect the output of production to assure that it meets the specifications. People do the inspection. This sets one person as judge over another's work, and because errors carry some sort of stigma, it has become normal procedure to ensure impartial inspection by making the quality control unit organizationally independent of the units whose work is being inspected. If this were not so, unacceptable products would inevitably slip into the finished goods.

However, functionally, inspection is only one of a long sequence of steps in the production of a finished part. The work may not proceed to the next manufacturing operation until the previous operation is demonstrated to be correctly done. We will regard QC in this book solely in this functional role, leaving the organizational problems of QC discipline to the factory manager.

An allied function, usually called Quality Assurance, is one element of the choice of a production technology. It is charged with the selection of a manufacturing technology which has the greatest probable yield of acceptable parts reaching the QC inspector. In other words, it anticipates errors and strives to prevent them, rather than detecting errors and ordering their correction. In this book we treat QA as a part of the manufacturing technology selection function.

Another good example of the difference between functional and organizational structure is Materials Handling. Because discrete parts manufacturing involves a sequence of operations on machine tools, which are fixed in location, it follows that the material being worked upon must move from tool to tool. This task usually falls to a unit in the factory equipped to lift pieces, or pans full of pieces, and move and deposit them elsewhere. This gross movement function is called for when the need arises, and gets done when the material handler fits it into his schedule. Materials handling is therefore a service function rendered as necessary; except for interplant moves, the cost and time for the moves is not separately accounted for. It is an "overhead" item.

Ask the Comptroller what it costs to move a thousand pounds of material a quarter of a mile in the factory, and you will get a shrug of the shoulders; no one ever asked that before, so the Comptroller has never looked into it. Ask how much it would be worth eliminating a dozen such moves a day, and the answer is: Nothing, unless you eliminate a whole unit—truck and man—in which case the savings can be calculated. (This line of thought does not apply to fixed conveyor belts or chutes, if they are used.)

However, functionally, a move – like an inspection – is one of a sequence of steps in the production of a part. The fact that it is performed by some one who reports through another channel of command is completely irrelevant for our purposes. In this book we will treat Material Handling as one of the functional elements in the production sequence.

## THE DIVERSITY OF THE PRODUCTION FUNCTION

The actual performance of the physical acts of production is an activity which differs very widely, depending on what is being produced, and in what quantities. We discussed the diversity of manufacturing in Chapter 1. It is probably safe to say that there are no two factories and no two production technologies that are the same. How then can one book be written that presents an understanding of manufacturing? It is my belief that there is a single fundamental set of principles which underlie all manufacturing regardless of what is being produced, and that is what I have tried to expound. It has been necessary to adopt the jargon and the environment of one industry in order to present the subject, and to leave to the readers from other industries the task of translating one set of terms into another.

Because of the tremendous size and the importance of the discrete parts industries to the gross national product, this book will use those industries as the source of its nomenclature. As discussed in Chapter 1, this definition encompasses all industries that procure bulk raw materials, fabricate a variety of individual (discrete) parts, and assemble them into the ultimate product for sale. While this may seem to exclude those industries that procure their input material from the extraction industries, and refine it into bulk materials, I believe that an examination of these "process" industries, with proper translation of the terminology, will show that the process industries also follow the same basic functional patterns.

It is my belief that this discussion is applicable, in principle, to any manufactured product, or to any process of manufacture. When one comes to adapt this discussion to a specific product or industry, it may be that some of the functions discussed here simply do not exist. For example, in a foundry there may be no assembly work whatever; in a computer plant there may be no part fabrication whatever; it is all assembly. In such cases, the diagrams of this book should still hold good, simply setting certain functions to zero, as it were, and the remainder will remain valid. As mentioned before, the idiom of each industry will vary, and changes in nomenclature will be required. This book has tried to stick to one meaning for each word. If the reader translates the words into his own language, he should take care to be consistent, and to retain but one meaning for each word.

Discrete parts industries follow one of four common patterns of operations. All share the common characteristic that the production consists of a number of operations identifiable as occurring at separate points in space and time. One pattern is the sequential pattern, in which a unit of material moves successively through a string of operations. This is the common pattern in shaping individual parts, and is even characteristic of whole procedures, such as cotton carding, spinning and spooling; or in newspaper printing, collating, cutting, and folding. Waste material, such as chips and dirt, may be extracted, and supplies, such as ink, may be added, without loss of generality. Figure 6.1 suggests this pattern.

Another pattern of operations is the combinative pattern, in which several parts are brought together and fastened in a predetermined orientation to one another. This is the common pattern in assembly operations, and is the definitive characteristic of the discrete parts industries. Usually small groups of parts are assembled together into subassemblies, several of which, with other parts, may be joined into larger subassemblies, and ultimately into the fully assembled end product. The fastening may be irreversible, as with riveting or welding, or reversible, as with nuts and bolts. The fastening may hold the parts in a fixed spatial relationship, as in putting the cylinder head on an automobile engine cylinder, or in moveable relationship, as in putting the pistons, connecting rods and crankshafts in those same automobile engines. The relative timing of the several operations is most important. Figure 6.2 suggests this pattern.

A third pattern of operations is the disjunctive pattern, in which a complex input is separated into several different components which

(Supplies)

(Waste Material)

**Figure 6.1.** Sequential pattern. Sequential operations may be performed on material, all of which moves through the process. Waste material, such as chips, may be extracted; and supplies, such as paint, may be added without loss of generality. The ordering of the operations in the sequence is ordained by the process.

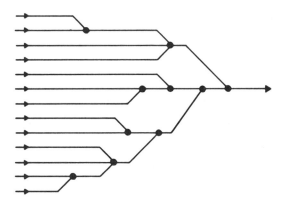

**Figure 6.2.** Combinative pattern. Assembly operations follow this pattern, although, frequently, they are much more complicated than this diagram would imply. Timing of operations is important.

constitute the several end products. An oil refinery, a slaughter house, or a saw mill would be examples. Because this pattern emphasizes the "many from one" mode rather than the "one from many" mode, it is rarely found in the discrete parts industries, but is the prototype pattern of the process industries. The relative timing of the various operations is unimportant. Figure 6.3 suggests this pattern.

Common parlance tends to divide all industries into "process" industries and "discrete parts industries," a classification which should be used with care. It is based on the criterion that process industries

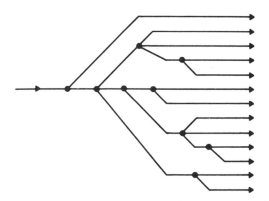

**Figure 6.3.** Disjunctive pattern. Naturally occurring raw materials frequently follow a disjunctive pattern in the initial stages of their processing. Usually the relative timing of the various operations is unimportant.

change the physical or chemical character of a material, whereas discrete parts industries change the geometrical form and dimensions of a material. The criterion is applicable to operations or groups of operations, but not properly to an industry. To illustrate: An automobile factory is surely a discrete parts industry, yet paint ovens and heat treating furnaces are process operations. A beer distillery is surely classified as a process industry, yet putting the caps on beer bottles is certainly a discrete parts operation. Actually, both of these types of industry use both "process" and "discrete" operations. Hence the potential for misunderstanding.

Another common parlance speaks of industries as "batch" or "continuous flow" industries. The terminology "batch parts manufacturing" is inappropriate if used to describe discrete parts manufacture. Many chemical products, coming from a process industry, are nevertheless made in batches, as are many metal products. On the other hand, many plants which are part of the process type industries use batch or discrete operations. It is well to avoid the term "batch" when describing discrete parts industries.

A fourth common pattern of operations is the locational pattern, characterized by the receipt, storage and disbursal of materials substantially without change, from some site, such as a warehouse. Figure 6.4 suggests three variants of the pattern. Incoming material may be received in small lots that are sorted into bins and accumulated to be disbursed later in large lots (sortation). Incoming material may be received in large lots to be held and dispensed in small lots (order picking). Or material may be received stored, and disbursed in the same sized lots at a later time (the familiar storage warehouse). It is interesting to note how often the locational patterns are present between each of the other three patterns.

## THE STRUCTURE OF THE PRODUCTION FUNCTION

This complex function of production is best addressed as a six-part activity fulfilling one of the four major functions of the whole manufacturing enterprise. The designs which Production accepts from the Development Function control what is to be made, and the directives from Manufacturing Management control when and how many products are to be made, and authorize the expenditure of time and resources. Production consumes input materials, uses resources, and produces finished

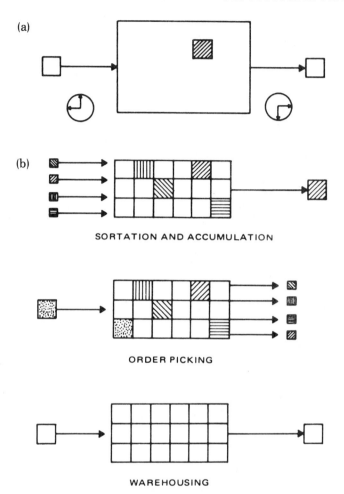

**Figure 6.4.** (a) Locational pattern. Receipt, storage, and disbursal are the important events, rather than alteration of the material. Timing is an essential element. (b) Variants of the locational pattern. Although any one of these three variations can be controlled either by a computer or manually, the mechanisms involved are different.

products and parts. Production's relationship to the other major functions is shown in Figure 3.2.

Also shown in Figure 3.2 are mechanism inputs to Box 3. These are the plant facilities, machinery, tooling, and skilled personnel which make production operations possible. Not shown are three other, more intangible, but nevertheless very important mechanisms which are present and active in all parts of the enterprise, and are therefore not repeated over and over again. These were discussed in Chapter 3: the backlog of experience and knowhow resident in the personnel, and implicit in the machinery and tooling; and the knowledge of the technology which is applied in the production function.

Figure 3.2 shows feedback from Production, Box 3, to Manufacturing Management, Box 1, of status reports and problems so that the administrative unit will know what is going on; and to Development, Box 2, telling of the manufacturing capabilities of the factory, so that they may design to use them most advantageously; and also forwarding requests for design changes to facilitate manufacture.

Turning inwardly now to the Production Function itself, Figure 6.5 displays the six-part structure; it is not surprising that in such a complex structure there are many interpart relationships. There are two distinguishable levels of planning, two types of procurement activities, an administrative unit to make schedules and budgets, and the material conversion unit. We will first examine these six functions which collectively are needed to produce outputs, and then we will devote a chapter to each. Because of this complexity, Figure 6.5 is broken down into two figures, called "FEO's" in the IDEF$_0$ terminology. FEO denotes "for explanation only," and does not supercede the original. The first FEO is Figure 9.1, in which all six function boxes are shown, but only the inputs and controls are traced. In the second FEO, Figure 9.2, the same six boxes are shown, but only the feedback flows of routine reports and problem reports are shown.

The reader will note that Figure 6.5, Produce Product, has the same name and the same inputs, controls, outputs, and mechanisms as does the A-3 Box of Figure 3.2. Figure 6.5 merely expands upon the details of the contents of Box 3. Later we will still further expand upon Box 6 of Figure 6.5, discovering details which could not be shown in the already crowded parent diagram.

As mentioned above, there are two quite distinguishable and quite essential planning functions involved in production. The first is called Planning for Manufacture (Box 1) and the second Planning for Production (Box 3). When the company's product is complex – automobiles, airplanes, or machine tools, for example – both levels of planning are present; for simple products they are still both required but are not so easily distinguished. When an individual part design is modified by Development, there may be no need for the Planning for Manufacture step, but there will be a need for Planning for Production.

Planning for Manufacture determines the strategy for producing the product. The major subdivisions of the product are identified. In

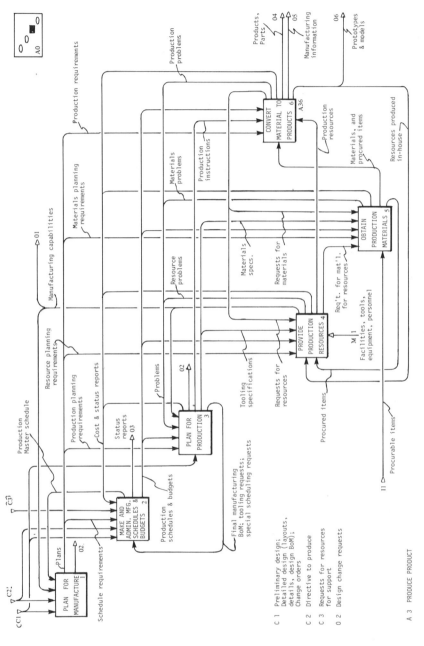

A 3 PRODUCE PRODUCT

C 1  Preliminary design;
     Detailed design (layouts,
     details, design BoM);
     Change orders

C 2  Directive to produce

C 3  Requests for resources
     for support

0 2  Design change requests

**Figure 6.5.** Produce products (A 3). Decomposition of the "Produce Products" box (Box 3) of diagram A 0 (Figure 3.2). The six-component functions produce a crowded diagram, reflecting the many and complex interactions of the six functions. To make this diagram easier to read, it is broken down into two "explanation" drawings, which are shown in Figures 9.1 and 9.2.

an airplane, there will be the fuselage, the wings, the engines, the controls, etc. Where are the interfaces between the parts? Will the parts all be made in one factory? If not, where will they be assembled? Will it be a fixed assembly area or a moving assembly line? Will the engines be made or purchased? What will be the manufacturing methods for the major parts? What are the probable times to produce, and the costs? What facilities are available in-house, and what must be procured? How long will that take? Support activities must be planned – quality assurance, materials procurement, tooling construction or procurement, facilities and equipment planned, and personnel readied.

Trade-offs will have to be made between the various alternatives, compromising costs against budgets, times against schedules, facilities available versus the time and cost to acquire more appropriate facilities. Personnel skills may be a problem. New technology may have to be sought out to replace available technologies which are not so modern nor cost effective. As a result, there will be many an iteration of the planning-for-manufacturing exercise before an optimum plan is fixed upon.

In contrast, Planning for Production is done to determine the method of production of individual parts, and the methods of assembling the parts into the end product. A complete manufacturing parts list or bill of materials will be written. Make or buy decisions will be made for each component, which may involve getting both in-house manufacturing costs and bids from sub-contractors. For each part to be made in-house an operation list will be written determining how it will be made and on what tools. Raw materials lists will be prepared. Assembly sequences will be written at the lowest subassembly levels and at all succeeding higher levels. The necessary machine tools, cutters, fixtures, and gauges are listed and availability determined. Cost and time estimates for all of this will be made and checked back against the allocated times and funds.

Schedules and budgets must be made and maintained to provide coordination of all the various Production activities. This function is shown in Box 2 of Figure 6.5, labeled "Make and Administer Schedules and Budgets." It operates under the control of the initial plan for manufacturing, taking into account the designs received from Development and the directive to Produce received from Manufacturing Management. It also receives the details of the Plan for Production, and out of this total body of information it generates a Master Schedule. This document is the road map for the whole Production operation. Needless to say, the Master Schedule itself is subject to revision as events unfold, but it is the basis for status and progress reports to all concerned.

There are two types of procurement functions involved in production – one for procuring materials and one for providing resources. Resources, it will be recalled, are things essential to production, but which are not physically incorporated in the products to be shipped out the door. Based on the plan for production, this function will determine what facilities and machinery will be required, what fixed tooling (jigs and fixtures, for example) and what expendable tools (cutters and drills, for example) will be needed, and what personnel will be required. When these resources are not available, this function will provide them. It is shown as Box 4 in Figure 6.5.

With the rapid evolution of production technology it sometimes becomes necessary to acquire new technology in order to produce competitively, or to assure product quality. Technology is an intangible, but the use of technology does require tangible resources. Search for and acquisition of this kind of resource is essential, and is properly allocated to this procurement function.

The second procurement function acquires the materials which will be converted into products, and also such materials as may be needed for in-house construction of tooling and equipment, or models and prototypes. This function is shown in Box 5 of Figure 6.5, labeled "Obtain Production Materials". Materials may be raw material to be converted into parts – metal sheets, bars, rods, or the like – or may be finished components manufactured by others – motors, hardware, hydraulic fittings, ball bearings, etc. – which will be included in the product during assembly. Procurable items will be purchased, received, inspected before acceptance, and stored until needed on the factory floor. Inventory control of all of this material is a major task.

The final element of the six-component elements of Production is the actual conversion of input materials into parts and their assembly into the finished products. This function is shown in Box 6 of Figure 6.5, labeled "Convert Material to Products." In the discrete parts industries this conversion takes the form of a sequence of events, following the patterns discussed in the previous section. Events may include moves in space, from point to point; or moves in time – storage from one time to a later time. Events may include changes in the shape or configuration of the material by cutting or forming, or changes in the physical character as by heat treating, plating, etc., and events may include inspection, assembly, and testing.

There may be many, many events in the production of a single part, and there may be many parts assembled into a product. Events occur in a fixed sequence, but the timing of each event at its particular facility presents a mammoth scheduling task. For any event, the following must be present at the machine tool: the material, the fixed tooling, the expendable tooling, the controlling instructions, and the operator. All of these have to be programmed and controlled.

In so complex a process, operating on so many individual parts, problems are to be expected. Budgets and schedules have built-in allowances for solving these problems. The solutions are as varied as the problems, and may require reworking parts or referral of the problem back upstream for changes in designs or specifications.

The final output of the Production function may be either products such as machinery or individual parts, such as hub caps. When packed and delivered to Marketing, they pass out of the domain of Manufacturing. Their distribution, installation, and support are the responsibility of the third of the four major functions of the total enterprise. A glance at Figure 2.1 will illustrate this point. However, the Manufacturing function does support the Marketing function by providing information, instruction and maintenance manuals, and necessary parts to keep the products working in the field. This support function is shown in Box 4 of Figure 3.2.

## SUMMARY

The Production function has six major functional subdivisions which, working together, accomplish the conversion of raw materials into finished products. These separate and distinct functions are not necessarily mirrored in the hierarchical organization of the production division of a company.

Production consists of a number of separate activities or operations occurring at separate points in space and time. Individual pieces are produced in a sequential pattern of events. Assembly follows a combinative pattern of events, a characteristic of the discrete parts industries. A locational pattern provides for storage of materials, and is frequently found between other events. In any given company it is probable that two or even all three of these patterns will be intermixed.

Of the six major functions comprising Production,

    a. Two are planning functions – one determining the gross structure of manufacturing, the other determining the fine structure of individual part creation and assembly;

    b. One is an administrative unit to make and administer schedules and budgets, and to create the Master Schedule;

    c. Two are procurement functions – one for resources needed

to perform the production, and one for materials consumed in production and converted into the final products; and

d.  The sixth is the conversion function which fabricates parts, inspects them, assembles them into products, and tests the products before shipment.

The output of Production is turned over to Marketing for distribution, installation, and support.

# Chapter 7

# Production: The Plan for Manufacturing

Planning for manufacturing is the first of the six parts of the production activity, which is shown in Box 1 of Figure 6.5, labeled "Plan for Manufacture." Its function is to create the overall strategy for production of the product, a task that is done but once for each product, even though the product may thereafter be produced many times. However, the strategy for manufacture may be revised from time to time.

The plans for manufacturing are based upon the directive from manufacturing management authorizing the production of products, shown as control arrow C2 in Figure 6.5. This directive will specify the number of units to be made, and the delivery schedule or rate insofar as it is known. The planning is also controlled by the product designs received from the development function, as described in Chapter 5.

If the designs have originated in the company development unit, then it is to be presumed that they conform to the established manufacturing practices of the company, and take cognizance of the manufacturing facilities available. On the other hand, the designs may have originated in some outside source. This would be the case in a job shop, or in any shop which was doing contract manufacturing for others. If it is merely a contract to produce X number of parts to a design drawing, then there is little strategy planning to do; the adaptation to local practice will be made in the "Plan for Production" function, Box 3 of Figure 6.5. However if the design calls for production of a complete product, then there may have to be some adaptation to the shop's facilities and skills. This would be done in conference with the buyer's design people, and possibly with the assistance of the local development group. It would be most wise to settle such adaptations before going too far into the work, preferably before the final contractual obligation is made.

If the designs are old familiar designs, merely being reordered, or with minor changes, then the manufacturing planning is routine. On the other hand the designs may pertain to a brand new product, and may not even be completely finished designs. This situation presents an opportunity for interdivisional cooperation for the ultimate good of the enterprise which is all too frequently overlooked or willfully foreclosed. For perspective please refer to Figure 3.2 and note the feedback of information on manufacturing capability from Box 3 (O1) to Box 2 (C3). In high technology industries, where there is rapid evolution of both product designs and of manufacturing technology, it is essential that this feedback line and the corresponding feed forward of designs from Box 2 (O4) to Box 3 (C1) be diligently and frequently used communication links. Note that an output from each of the functions acts as a control upon the other. The most successful companies actually station a person from Production in the Development area to ensure the feedback, and a person from Development in the Production area to interpret designs and act as feed forward liaison.

Input to the planning function may be of three sorts. The preliminary design (from Box 2, Figure 5.1) may be sent forward before the detailing is completed, in order to get advice on the preferences of Production on the design configuration, as suggested above. It makes little sense to design a part in such a manner as to make it difficult to produce. A delightful bit of doggerel, author unknown but common in drafting rooms 50 years ago, illustrates this point; it is reprinted in the following poem.

## THE SUCCESSFUL DESIGNER

> The designer bent across his board,
> Wonderful things in his head were stored,
> And he said as he rubbed his throbbing bean,
> "How can I make this hard to machine?"

> "If this part here were only straight,
> I'm sure the thing would work first rate,
> But 'twould be so easy to turn and bore
> It would never make the machinists sore.

"I'd better put in a right angle there
Then watch those babies tear their hair.
Now I'll put the holes that hold the cap
'Way down here where they're hard to tap.

"Now this piece won't work, I'll bet a buck,
For it can't be held in a shoe or chuck;
It can't be drilled or it can't be ground,
In fact the design is exceedingly sound."

He looked again and cried, "At last —
Success is mine, it can't even be cast!"

Early information on new designs may be very essential to permit Production to order some items with very long procurement lead times. A good example of this is the engines for innovative aircraft designs. They are usually made by a company that specializes in the manufacture of engines, and not by those who make the airframes. Lead times for development and production may be so long that it would be poor judgment to delay the engine purchase until the design is finally complete. Aircraft people tell me that the preliminary design department orders the engine, not production; this merely illustrates the difference between purchasing as a function and purchasing as an organizational prerogative.

The detailed design, when completed, is transmitted to the group planning manufacturing strategy. This design will consist of layout drawings or their equivalent, a detail drawing for every part to be made and a specification for parts that are to be purchased as they will be used, and the bill of materials. The detailed design governs, in case there is a difference between it and the preliminary design previously received.

Another type of input design is the modification of an existing part drawing or layout. These are incorporated in what is frequently called an Engineering Change Notice, or Engineering Change Order (ECN or ECO). The notices result from requests from later stages of production or requests from the fields for modifications for any number of valid reasons. When development agrees that the request is justifiable and that it will not jeopardize the functional capability of the product, it makes official changes in the design and sends the new drawings and bills of materials on to production. Usually these are simply recorded and forwarded to scheduling and production planning for execution. Occasionally an engineering change may be substantial enough to merit some manufacturing planning review. However, from an administrative functional point of view, each change notice, large or small, is processed through the same steps as a whole new product.

Planning for manufacturing begins with an analysis of the product design. First, prior experience will suggest the optimum breakdown of the total product into its major parts. For example, a typical machine tool might consist of a base, a work table with its supporting slides, a column with a vertical slide, a spindle housing, and a motor and drive train. Usually such major parts may be separately assembled as units and then joined in the final assembly stage. In turn, the major parts may consist of logical subunits. When the product is a large structure, such as an airplane or a newspaper printing press, this breakdown may be carried down several levels without getting to the point of naming individual pieces.

Each of these major parts is considered to determine whether it will be made or bought. Electric motors are obvious candidates for purchase; gear boxes or ball bearing lead screws may well be acquired ready made from specialists. Major individual parts such as the rolls in a steel rolling mill may be purchased from someone with the very large lathe equipment necessary to turn such large objects. In general, make or buy decisions at this level of planning are not made on the basis of relative costs, but on the basis of the availability of in-house capability to make the parts. This may in turn raise the strategic question of whether or not to acquire additional capabilities. If there is enough work in prospect requiring new equipment or new technology, or even a whole new building facility, then the election to make or buy becomes a large scale economic decision. I recall one part design – the housing for a nuclear reactor – which was so tall that a new vertical boring mill had to be ordered. This in turn was so tall that a new building had to be constructed to house it and the cranes that lifted the parts.

The alternative to such facility changes is to request the design group to consider a modification of the design to permit the available facilities to be used. This is one of the many sources of requests fed back to the development group for design changes.

Thus a potential manufacturing strategy is devised; it is then evaluated on the basis of (1) the facilities available or purchasable, including therein the tools, test equipment, and manpower; (2) the cost of materials, of fabrication, of inspection, of assembly and test; and (3) the time required for start-up and to produce batches of products after start-up. These estimates are then fed back to the previous step for rethinking, until a satisfactory strategy is achieved. Each iteration of this procedure tests the manufacturing plan conceived against the constraints of time and budget set forth in the directive to produce which was received from manufacturing management.

Once a satisfactory manufacturing plan is adopted, it is then expanded and developed in greater detail. The sequences of part production and subassembly in batch mode, and of final assembly in line mode is planned. Part production times are roughly estimated, not on the basis of calculations from operation sheets (because these have not yet been prepared), but on the basis of prior experience. All the parts for a subassembly must be made before the subassembly work may be started. All the subassemblies must be done before final assembly may be started. Assembly times are roughly estimated, again on the basis of prior experience.

Putting all these time estimates in proper sequence leads to an overall schedule estimate, which must be consistent with the directive to produce. If not, restructuring and replanning must be done. It is conceivable, of course, that no amount of restructuring and replanning will fit the necessary work into the time available. In that case, manufacturing management is notified so that they may reconsider their directive or take other appropriate action.

In parallel with the development of manufacturing plans, supporting activities are also planned. This includes the following:

Quality assurance selects the technology and the equipment with the highest level of acceptable output achievable, consistent with costs and times allowed.

Materials planning identifies purchased items, including items with particularly long lead times; it identifies vendors and subcontractors, and probable material availability vs. time.

Tooling planning determines what kind and number of tooling items will be required, and which may be made in-house vs. purchased; also the needed hand held tools.

Facilities and equipment requirements are listed, and plans made to provide these resources.

Personnel requirements are compared to available skills and gaps are noted; hiring and training are planned. (This need may have a very long lead time.)

Snags in any of the above vis-a-vis the mandated schedules and budgets are noted and fed back, requesting appropriate change or guidance.

Referring now to Figure 6.5, we see this function of planning for manufacturing in Box 1. It has two outputs shown. First, the distribution of the overall plan as described above to each of the other functions within Production. It forwards the plan plus the time and money limitations to the function which makes and administers schedules and budgets, shown in Box 2. It forwards the plan with part production requirements to the function which plans production, shown in Box 3. It forwards the plan with resource requirements to the function which provides resources, shown in Box 4. It forwards the plan with materials requirements to the function which obtains materials, shown in Box 5. Finally, it forwards the plan with production time and quantity requirements to the shop function shown in Box 6 which will convert materials into products.

Second, it refers requests for design changes to the detailed design unit. These may have been referred to it by other units downstream, or may originate as the result of the planning function. For example, it may turn out that some type of material, or some piece of equipment or tooling, that had been assumed at the time of design to be available, is actually not available or is inadequate; or, when production of a part is viewed in the context of the total manufacturing activity, it may be possible to effect some cost or time savings if it were to be changed in some way.

## SUMMARY

Planning for manufacturing is the first of six functions which comprise Production. It receives product designs – for new products or revised products – from Development, or from an outside source in the case of job shops, and creates the overall plan for manufacturing.

The preliminary design of a new product may be sent ahead to production before the design detailing is completed, in order to forewarn production of any problems which might arise due to unusually long lead time procurement problems in materials, components, equipment or technological skills.

The product design is analyzed to determine the major structural divisions of the product; each major part is analyzed to determine whether it will be made or bought, based on the available capabilities. The question of adding a new capability is considered.

The manufacturing strategy is evaluated on capability, costs and time, and adjusted to conform to the constraints imposed by the directive to produce.

The selected manufacturing plan is then expanded and refined; a time schedule for manufacturing is created, based on past experience, to achieve the overall plan. Plans are also started for quality assurance, materials acquisition, tooling, facilities, and personnel as necessary.

The plan is distributed to the other five production functions for their guidance.

# Chapter 8

# Production: Budgets and Schedules

The making and administration of budgets and schedules is the second of the six functional parts of Production. Its function is to convert the strategic plan for manufacturing into time schedules and funds budgets, and to monitor and control the succeeding multiplicity of activities so that they stay within the constraints imposed by the directive to produce.

Control has been defined as "regulation by comparison to a standard." In the terms of this function, the standard is the set of schedules and budgets derived from the strategic plan which is received from the previous function, Plan for Manufacture. The comparison of actual performance to this standard is the monitoring function. What follows is the control, by noting deviations from the standard and taking, or permitting management to take the necessary actions.

An obvious necessity for control is a closed loop of feedback of information, and rapid communication with all parties. At lower control levels the control system deals with detailed information and short time horizons, where fast response is required. At progressively higher levels, information tends to become less detailed and the control-time horizons become longer with less rapid response being required.

The first and possibly the most important part of this function is the creation of what is known as the Production Master Schedule. This is derived from the strategic plan received from the planning function discussed in Chapter 7. Referring again to Figure 6.5, or the slightly less complicated version, Figure 9.1, this input plan with its time and money constraints is shown as arrow C3 to Box 2. The function is also controlled by the designs and design changes, shown as arrow C1, and the directive to produce, shown as arrow C2 entering Box 2. A further control is the requests for resources for product support; should spare parts or service tools for use in the field be required, these items will be added to the production schedule. This control is best seen in Figure 3.2.

Time spans for all parts of the plan are calculated. This includes the time spans for procurement of materials and resources, the time spans for fabrication of parts, and the time spans for assembly and test. Some of these time spans may run concurrently, for example, several parts may be in the process of fabrication at the same time. Other time spans must run sequentially, for example, procurement time must precede fabrication time, which in turn must precede assembly time. When there is a conflict between two parts that are seeking simultaneous access to a machine tool or other facility, the conflicts must be resolved by trade-offs involving costs vs. scheduled time or costs vs. performance. Unresolved conflicts are referred back to higher authority for resolution.

Initially a Production Master Schedule is produced from the exercise described above, and is published for the guidance of all concerned. It goes back to the Plan for Manufacture function, and downstream to all the other four functions in production. Figure 9.1 shows this distribution downstream, and Figure 9.2 shows the distribution upstream. Any subsequent changes in the schedule due to slippage or to negotiated changes are duly incorporated in the Master Schedule. Any changes in the production requirements or in revised manufacturing plans will also require replanning and inclusion in the Master Schedule.

All elements of the Master Schedule are combined to determine the flow time through the process of manufacturing for the first article. This gives a clue as to the possible start time for delivery. Flow times for successive batches are reduced in accordance with experienced learning rates and milestones for succeeding batches are reset accordingly. When milestones are reset, resource and materials procurement functions are advised.

Note that only those schedules that coordinate activities in more than one function area are handled here. There is further detailed local scheduling in materials control and production control which does not appear in the Master Schedule.

Cost estimates and budgets are set up to control cost, just as schedules are set up to control time. For convenience, the cost accounting packages are set up corresponding to the elements of the work plan. Each has a clearly defined work scope, to which costs can be related. Total costs for a project must lie within the assigned total budget, and problems must be resolved as before by trade-offs with time and performance. Unresolved cost overruns are referred back to higher authority for action or guidance. Initially cost estimates must rely on previous experience with similar manufacturing problems, but after the next function – planning for production – has done its work the cost estimating becomes more and more precise.

Costs in this context include all the expenditures necessary to accomplish the work. Labor and materials immediately come to mind, but supervision, maintenance, and so-called overhead labor, plus a charge for plant and machinery must be included. Costs also include the procurement of materials and components purchased from others, and the costs of subcontracted work being done for the company in outside plants.

Administering the schedules and budgets is the second major task of this function. It consists of monitoring activities and comparing actuals with plans, taking corrective action where needed. Monitoring methods are well known data collection, data consolidation, and data display techniques. Each enterprise will develop its own system. A wide variety of electronic data collection systems are available on the market. Some read directly from the machines, others require manual input at some specified intervals or at some designated milestones. There are very sophisticated computer systems to consolidate the collected information, display it in numerical or graphical form, and compare it to the schedules or budgets.

Undesirable discrepancies call for corrective action, usually by others than those who do the monitoring. If there is a project manager for the problem work, the corrective action devolves upon him. Otherwise the manager of the Production function is the likely candidate.

Engineering change orders, coming down through the system from Development have to be worked into the schedule and budget. Schedules are usually set back by these changes, unless work has not yet started on the part in question. Budgets may be affected either favorably or unfavorably by ECO's; the change could eliminate a part, or an operation, or substitute an easier or less costly method.

The next of the six functions in Production is called Planning for Production. It prepares, among many other documents, the final manufacturing bill of materials. It is fed back to the scheduling and budgeting function for incorporation in the Production Master Schedule and all other planning documents. This feedback is seen in Figure 9.2. Also seen there is the feedback from Boxes 3, 4, 5, and 6 showing their cost and status reports. These are data transmission systems.

The output of the function "Make and Administer Schedules and Budgets," which we have been discussing in this chapter, is most easily seen in Figure 9.1. The arrow leads to all four of the remaining boxes on the chart, becoming a control to each of them. Other outputs are, of course, reports upstream of status and the Master Schedule distribution, Figure 9.2.

## SUMMARY

Making and administering schedules and budgets is the second of the six functions which comprise Production. It converts the strategic plan for manufacturing into time schedules and budgets for funds, and then monitors the performance of the projects as they proceed through production, and controls the multiplicity of activities to conform to the plan.

It creates a Production Master Schedule showing the time spans of all the elements of the production process, including procurement, fabrication, and assembly. The Production Master Schedule is the guide for all the other functions in Production; it determines the flow time through manufacturing for the first article of each project. Flow times are adjusted in accordance with learning rates.

In the same manner, cost estimates and budgets are set up for each project, and costs are collected as work progresses.

The function monitors the progress of work, both in time and in cost, and relates progress to the schedules and budgets set up in advance. Corrective administrative action is initiated through the proper agents when necessary. Schedules and budgets are maintained up to date, and status reports are rendered regularly.

# Chapter 9

# Production: Planning for Production

Planning for production is the third of the six parts of the Production activity. It is shown in Box 3 fo Figure 6.5, labeled "Plan for Production." Its function is to convert the product design, as it comes from Development in the form of detailed drawings and specifications, into detailed manufacturing methods and instructions. These instructions must conform to the overall plan for manufacturing, which was generated in the first of the six production functions, and to the schedules and budgets which were generated in the second of the production functions. These plans and schedules are shown in Figure 6.5, or more clearly in Figure 9.1, as controls to the production planning function.

In turn, Planning for Production sends production instructions to the shop, which is shown in Box 6 of Figure 6.5 and labeled "Convert Material to Products." Planning for Production also sends tooling specifications to the function that provides production resources, shown in Box 4, and materials specifications to the function that obtains production materials, shown in Box 5. In the course of all this detailed planning, this function will generate what is known as the final bill of materials. There was, of course, a bill of materials which came down from Development with the complete product design, but Production may justifiably make some modifications in it, and may request some minor

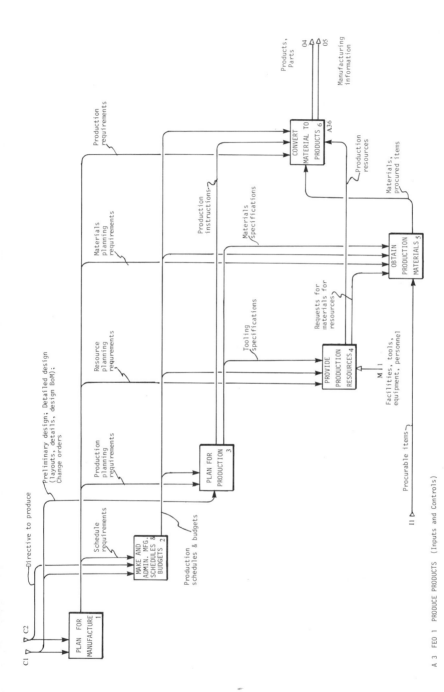

A 3  FEO 1  PRODUCE PRODUCTS  (Inputs and Controls)

**Figure 9.1.** Produce products (FEO 1 of A 3). The same six functions shown in Figure 6.5 are shown here, with only the data paths (arrows) which carry inputs and controls. The heavier lines show the principal paths of incoming designs and materials, and their delivery as finished products.

changes in part designs to facilitate manufacture. The redesign is checked with Development, and the final bill of materials and any new tooling or special scheduling requests are fed back to Scheduling and Budgeting, Box 2, for their information and administrative purposes. These feedbacks are more clearly seen in Figure 9.2. It will be recalled that Figures 9.1 and 9.2 are extracts from the complete diagram of Production, shown in Figure 6.5. They are "For Explanation Only" and do not supercede the complete diagram.

It is at this stage of manufacturing that the process seems to explode into a colossal number of minute details. Management will look upon a project as a single entity; Development will look upon a design as a single entity, composed of several parts; Planning for Manufacture deals in the larger subassemblies, as does Scheduling and Budgeting, even though each part of a product has to be considered. However when it comes to the Production Planning function, not only every single part, nut, bolt and pin has to be kept track of, but every operation on every part has to have machines, tooling, gauges, and supplies planned out in advance, ordered, procured, delivered, and scheduled.

## FABRICATION PLANNING

**First, for each element of the product design, a detailed manufacturing method is devised.** This means every part of a product, whether large or small, single or multiple. The design bill of materials is the primary source of the list of elements to be dealt with. It will probably not list, however, such things as paint, lubricants, metals to be plated on to some parts, hydraulic fluids, adhesives, special packing materials, and the like. They may be noted on the detailed drawings, but not as component elements. It is easy to see why the final bill of materials may be larger than the design bill of materials.

For each item on the list, the make-or-buy decision comes next. Some parts are obviously going to be made in-house, if the company is a manufacturing enterprise. Some parts are equally obviously best procured from others who specialize in them—hardware, electric motors, ball bearings, and the like. However some parts may present an option. Even if the capability exists to make the parts in-house, they may still be subcontracted to a job shop or a specialty manufacturing company. The determinant may be economy—which will be cheaper? The determinant may be scheduling—can an alternate source deliver the parts on the scheduled date in the needed quantities? Or it may be strategically desirable to make a few parts in-house and farm out the majority of the

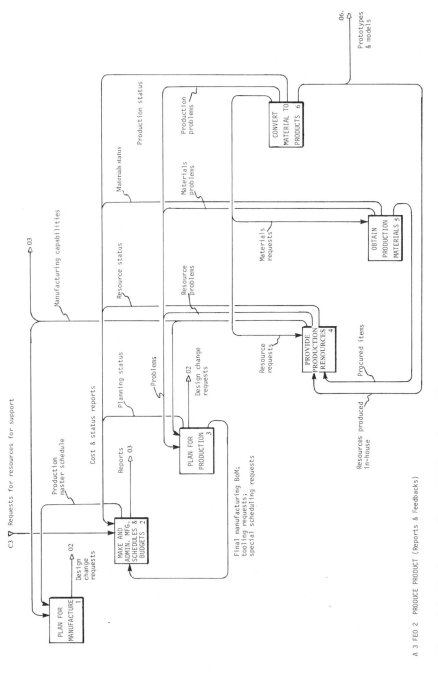

**Figure 9.2.** Produce products (FEO 2 of A 3). The same six functions shown in Figure 6.5 are shown here, with only the data paths (arrows) which carry reports and feedbacks. Of note is the feedback of information on manufacturing capabilities available in the factory, for the guidance of planners, designers, and managers.

requirements, thus retaining the capability in readiness in case the outside source should be cut off some day; or to have in-house experience as a yardstick with which to evaluate the subcontractor's prices and his reported problems.

If the make or buy decision is to be made, then the part under consideration must be put through production planning in order to estimate the in-house cost, as well as offered to outsiders to estimate and bid competitively.

Actual procurement of those parts which are selected for purchase will be conducted by the Materials Procurement Function, Box 5, because they are specifically equipped to handle such functions, but the number and delivery dates, the specifications, and the price will be as established by production planning.

Economy in manufacture may suggest that the end form of a "part" to be made, in the context of the present discussion, may not be its ultimate end form. For example, shafts for small electric motors may be basically identical in most dimensions, differing from model to model only in the configuration of the ends of the shafts. One buyer may want the end threaded to receive a nut; another may want a keyway as well as a threaded end; still another may want one (or two) flats for a set screw. The motor producer can make shafts economically in large batches, leaving the ends large enough for any variation of design. Such parts are called "semifinished" parts. This intermediate stage, with the generic end shape, is treated in planning as if it were a finished part. Parts will be stocked but never incorporated in a finished motor until each has been given its final end shape to suit the specific end user. They will then carry new identifying numbers.

For another example of another type, an auto engine connecting rod and its cap may be made as two distinct parts. They are then mated and together are further machined before they are ready for assembly into the engine. For purposes of production planning the connecting rod and the cap are treated as two separate parts before mating, and as a third and separate part after mating. Thus we see that production planning may for its own purposes create new part identities that did not exist in the designer's bill of materials, and may do so without in any way changing the engineering design.

**Second, for each part that is to be manufactured in the plant, a sequence of operations is listed showing every act that is necessary to convert a raw material input into a finished part.** Such documents have many names: route sheets, operations lists, process sheets, methods planning sheets, and many more. The actual terminology used is not important; it will follow local custom. In theory they should contain every act or event necessary, including operations which change the shape or dimensions of the part, its surface finish or materials characteristics (e.g., hardness), verify its accuracy or performance, and specify its movement or storage; they should list the machinery, tooling and skills necessary for each act or event. In practice much of this information is unwritten, relying instead upon the knowledge of the operators and the shop practice standards established by long usage.

Certain information is absolutely essential to operations lists: part name and number, the material from which the part is to be made, the size of the raw material; and a list of the major machining operations to be used, in sequence. Usually such brief documents are accompanied by a copy of the detailed drawing. A sample of a brief operation sheet is given in Figure 9.3, and the drawing of the part to which it pertains is given in Figure 9.4. In many companies this would be a perfectly adequate pair of documents to achieve production of this part. As pointed out in Chapter 5, the detailed drawing should contain all the information necessary to creation of the part so that it will perform its allotted function. The operation sheet indicates the succession of operations necessary to convert the raw material into the part shown on the drawing.

Creation of the operations list involves the skill of an operator who is completely familiar with the production technology of the shop and the machine facilities available. The operator writes the sequence as he deems it best, adding details for the guidance of the shop as he believes necessary. It can be seen that the generation of operation lists in a busy shop could be a tremendous task. Recently computers have been applied to assist in the work, and the operation is then known as computer aided process planning (CAPP). Several such systems have been developed by commercial vendors, academic institutions, and one which was sponsored by the trade association in this field, Computer Aided Manufacturing, International, Inc.

There are two types of process planning systems, the first to be developed is known as the variant system, and the more recently developed as the generative system. The variant system develops operations lists for new parts by making variations on proven plans used on similar parts in the past. The generative system develops operations lists for each part based upon its configuration, using computer algorithms to convert the geometrical and dimensional data in the detailed part design into a sequence of operational steps.

Use of the variant system requires either an almost infinite and precise memory on the part of a human planner, or the recall of recorded plans by a system which can identify reasonably relevant data on file.

| OPERATION LIST | | | | |
|---|---|---|---|---|
| PART NAME<br>Mandrel | | PART NUMBER<br>12345 | MATERIAL<br>Mild steel 1" rd. | |
| OP | DEPT. | OPERATION | | |
| 10 | 01 | Cut off to length | | |
| 20 | 16 | Turn and bore | | |
| 30 | 14 | Drill and tap 2 holes | | |
| 40 | 22 | Deburr, stamp number | | |
| 50 | 90 | Inspect | | |
| 60 | 75 | Store | | |

**Figure 9.3.** Operations list (short form). More commonly found operations list, containing only the major machining operations.

This, in turn, requires the use of a coding and classification system to identify these sources of past experience. Each newly designed part is classified by its physical attributes, as discussed in Chapter 5. The code number can be used to retrieve the operations lists of several previously manufactured similar parts. The planner selects the best fit and then modifies it as necessary to suit the part in hand at the moment. The

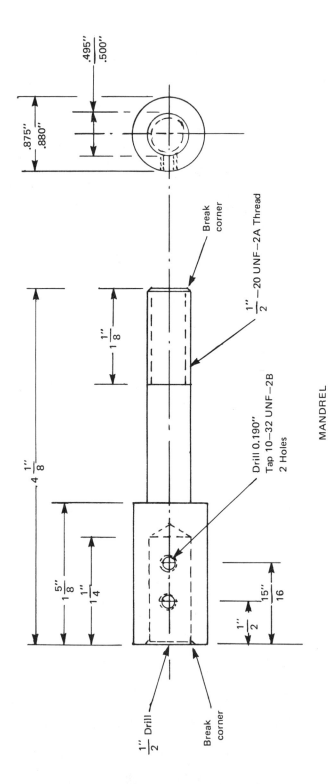

MANDREL

**Figure 9.4.** Detail drawing for operations list of Figure 9.3.

computer's word processing capabilities allows him to produce a written operation list with a minimum of new writing. This procedure is substantially what was done before computer recall was available, but it is now many times faster and more useful.

There are some criticisms of the variant system of process planning. To use it, the backlog of experience must be classified and retrievable; this requires an enormous amount of work when the system is first introduced. The system tends to repeat old technology, right or wrong, rather than encouraging the use of new tools and technology.

The generative planning systems now being developed (1983) by Boeing, Lockheed, and others rely upon a complex computer program which accepts data from a CAD computer's data base and specifies the necessary operations in the correct sequence to generate the required end product. There is little human input except to make an occasional choice between two alternatives, or to intervene when a problem arises which the computer has not yet been programmed to handle. New technology, or new facilities, may be added to the computer software and will then be used by the system when next the occasion arises for their use.

Either by human brain power, or by the variant or the generative computer systems supervised and abetted by human skill, an operations list must be produced for each part.

The operation list, in addition to its many other functions, acts as a routing guide for the movement of the material through the shop from point to point. A copy of the operation sheet accompanies each batch of material on its journey.

A work order, telling how many parts are to be made in this batch, and the account to which the work is to be charged, is either added to the operation list and the detailed drawing as a third document, or included in the operation sheet. Combining them makes the operation sheet uniquely associated with a specific batch of production; if the work order is kept separate, the operation sheet may be reused for subsequent batches until the design is changed.

The operations list serves as a basis on which to predict times and costs. Virtually everything that takes place in manufacturing takes time and creates costs, and the more detailed information the estimators have available, the better their estimates will be. Therefore if an operation sequence listed every act or event, and every mechanism required, it would serve better as an administrative tool. Figure 9.5 shows a much more complete form of operations list. It contains:

1. The name and number of the part;
2. The input material and the size of that material;
3. Every material cutting, forming, surface treatment or heat

# OPERATIONS LIST

| PART NAME | PART NUMBER | MATERIAL | SIZE |
| Mandrel | 12345 | Mild steel | 1" Round |

| OP | SER | DEPT. | OPERATION | MACHINE | TOOLING | SET-UP | CYCLE | SKILL | REMARKS |
|----|-----|-------|-----------|---------|---------|--------|-------|-------|---------|
| 5 | 0 | 01 | Bar stock from storage | | Tote Pan | | 4 | 2 | Rack 15-2-A |
| 10 | 0 | 01 | Cut off to rough length | Power hacksaw | | 1 | 1 | 1 | |
| 15 | 0 | 82 | Move to Lathe Dept. 16 | Truck | | | 8 | 1 | |
| 20 | 0 | 16 | Face end | Chuck. Lathe 869 | 419F | 5 | 1 | 4 | |
| | 10 | | Rough & Finish turn lg. OD | | 429T | | 3 | | |
| | 20 | | Drill end | | $\frac{1}{2}$" drill | | 2 | | |
| | 30 | | Rechuck | | | 1 | 1 | | |
| | 40 | | Face end | | 419F | | 3 | | |
| | 50 | | Rough & finish turn sm. OD | | 429T | | 4 | | |
| | 60 | | Thread | | VTh. 16 | 2 | 4 | | |
| 25 | 0 | 82 | Move to Drill Dept. 14 | Truck | Tote pan | | 5 | 1 | |
| 30 | 0 | 14 | Drill 2 holes | Drill press G55 | 0.019 drill | 3 | 3 | 3 | |
| | 10 | | Tap 2 holes | | 10-32 tap | | 4 | | |
| 35 | 0 | 82 | Move to Bench Dept. 22 | Truck | Tote pan | | 4 | 1 | |
| 40 | 0 | 22 | Deburr, stamp number | Hand | Hand tools | | 3 | 2 | |
| 45 | 0 | 82 | Move to Inspection 90 | Truck | Tote Pan | | 3 | 1 | |
| 50 | 0 | 90 | Inspect | | Hand tools | | 2 | 3 | |
| 55 | 0 | 82 | Move to Finished Stock 75 | Truck | Tote Pan | | 4 | 1 | |
| 60 | 0 | 75 | Store Bin 678 | | Tote Pan | 2 | 2 | 2 | Hold until needed |

**Figure 9.5.** Operations list (long form). Operations list containing everything that will be done to the material, including moves and storages. Tools are listed with setup time, cycle time, and operator skill required. Some operations, such as No. 20, are broken down into suboperations performed sequentially on a single machine tool.

treatment operation that changes the configuration of the part;

4. Every inspection or testing operation;
5. Each machine tool used and its location;
6. Special fixtures, clamps, gauges, etc., employed;
7. Special cutting tools or form tools employed;
8. Supplies such as paint, oil, pallets, etc.;
9. Every move from point to point in the factory;
10. Every storage location and time dwell, excepting time spent in queues at machines;
11. Skills used at each operation, plus helpers as needed.

Except for No. 1 in this list, every item is a source of an assignable cost or costs, and an estimated time span. Every item will be used to schedule events, call out tools and supplies as needed, and generally tighten up the administration of the shop. In closely controlled enterprises, such operations sheets are essential.

There may be but one or two operations needed to complete a simple part, but for ordinary parts it is not uncommon to have a dozen or more operations on the list, and complex parts may require many more. When a batch of parts are to be processed together, the entire batch is worked upon at each machine before the batch moves to the next operation.

**Third, for each operation on each part, the methods and resources are listed and described.** For each operation there must be present:

1. The work piece or pieces;
2. The machine tool or its equivalent which is to do the work;
3. Operator or operators with appropriate skills;
4. All necessary expendable tooling – cutters, drills, abrasive wheels, etc.;
5. All necessary nonexpendable tooling – chucks, clamps, fixtures, jigs, templates, NC tapes, etc.;
6. All necessary materials handling adjuncts – pallets, cradles, tote boxes, etc.;
7. Documentation, including the work order, the operations or route sheet, the instruction sheet, and the detailed part drawing and set-up diagrams.

Numbers 2 and 3, the machine tool and its operator, are at a fixed location in the plant. Numbers 1 and 7 and probably the NC tapes will accompany the parts to that site in accordance with the information on the route sheet. All the rest will be called out from various storage areas and delivered directly to the work site. A sample instruction sheet and set-up diagram are shown in Figure 9.6.

The information necessary to trigger and control all these moves, without which work cannot be done, is contained in the operations list and instruction sheets. If it is not explicitly spelled out there, then the company must rely on the memory of the staff to follow routine procedures in place of the written information. It is one of the key duties of the Production Planning function to provide this information to the extent needed.

Further, the planners can calculate the time necessary to accomplish each of the elementary steps in the operation. Files will provide records of previous experience, or motion–time analysis can compute the times when no experience exists. The aggregation of the many time elements will be translatable into total time for each operation, and this in turn will translate into costs for cost accounting.

In most manufacturing operations there will be a set-up time during which the operator mounts the holding fixtures on the machine tool, places the part in the fixture, places the cutting or forming tools in the machine tool spindle, puts the NC tape or template in place, establishes axis datum points, and checks out the set-up. A foreman's approval may be required on important work before the first piece is operated upon. The machine's work cycle time will be repeated for every part in the batch, if there are more than one. When the last part is done, the operator breaks down the set-up, prepares the cutters, fixtures, etc., for return to their storage areas, prepares the batch of parts for movement to the next machine site, cleans up his own machine site, and completes the recording of the completed job. The latter will, when transmitted back to production control, trigger another whole sequence for the next operation on the route sheet.

The work cycle time and the set-up and breakdown times allowed may be shown on the operations lists or on the instruction sheets, and have to be included in the total time calculations referred to above. If move times and storage times are known and can be added in, a good approximation can be made of the minimum throughput time for a batch of parts. This time would indeed be the true time if there were no delays at any operation site, waiting for the machine to become available. In emergencies, a batch may be expedited through the plant by setting aside all other work which might cause delays. However in ordinary circumstances, a queue of jobs are lined up at each machine, each with its assigned priority. This queue allows the foreman to juggle work sequences to adjust for unforeseen delays such as operator absence, tool breakdown, power failures, and other emergencies. Production planners usually have a rule of thumb for how much of a queue to provide as a

| NC Combined Tool and Set-up Sheet, KT4 | | | | | | |
|---|---|---|---|---|---|---|
| Part Name *Support Plate* | | | Op. No. *050* | Fixt. No. *1063880* | | Part No. *588235* |
| Remarks | | | Mach. Group *1704* | Tape No. *1057137* | | EC No. *719 Y* |
| GR | No. | Tool Description | Diam. | Length | Pocket | Notes |
| *04* | *18* | *Spot drill* | *0.188* | *7.500* | *1* | *1" Max. tool* |
| *26* | *16* | *Drill* | *0.201* | *6.500* | *2* | *extension on all tools* |
| *30* | *00* | *Tap (1/4-20)* | *0.250* | *7.500* | *3* | *beyond tool holder* |
| *02* | *21* | *Drill* | *0.136* | *7.500* | *4* | |
| *30* | *13* | *Tap (8-32)* | *0.164* | *7.500* | *5* | |
| *20* | *12* | *Drill* | *0.180* | *7.500* | *6* | |
| *05* | *30* | *Ream* | *0.1869* | *6.500* | *7* | |
| *08* | *13* | *Drill* | *0.159* | *7.500* | *8* | |
| *30* | *02* | *Tap (10-32)* | *0.190* | *7.500* | *9* | |
| *09* | *15* | *Drill* | *0.1094* | *7.500* | *10* | |
| *30* | *05* | *Tap (6-32)* | *0.136* | *7.500* | *11* | |
| | | | | | | |

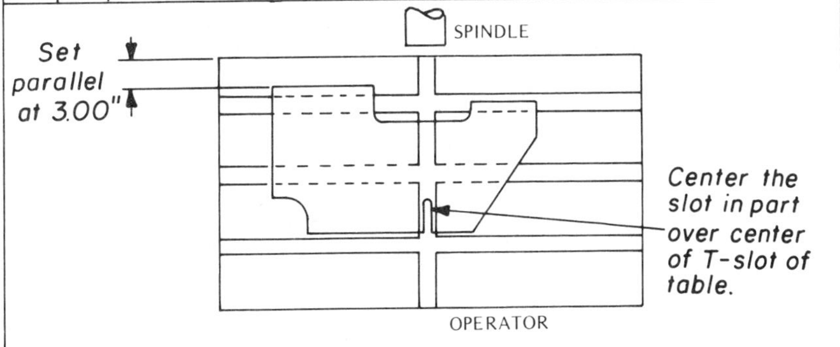

*Set parallel at 3.00"*

SPINDLE

*Center the slot in part over center of T-slot of table.*

OPERATOR

**Figure 9.6.** Instruction sheet and setup diagram. Type of information supplied to a machine operator—in this case an NC machining center. Tool designations, sizes, and lengths from tip to datum point on the tool holder, and their location in the tool magazine are shown. Also placement of the work piece on the work table.

cushion for these minor glitches. They may schedule at the rate of one operation per day, or per two days, or per week, depending on how good their shop control is. A more detailed discussion of this scheduling problem will be found in Chapter 12.

For items which are to be bought, rather than made, the production planners will prepare the drawing of the part or the specifications in such detail as will assure that, when delivered, the part will be what the designer wanted. They will set up the correct receiving inspection criteria, so that the receiving clerk will know if incoming inspection is required, and if so, to whom the part should be sent for that function. If accepted by the inspector, storage instructions must be provided so that the incoming material will be safe and available when needed for assembly.

## ASSEMBLY PLANNING

Next, assembly operations are planned, and the necessary tools and equipment are listed. There are very few products which are not comprised of several parts, fabricated and then assembled together to make the whole. Even the common lead pencil has four parts. Coins, paper clips, house keys, and teaspoons are among the few exceptions to this generality.

Assembly proceeds in a reverse tree pattern. A few parts are assembled into a subassembly. Other sets of parts form other subassemblies. Subassemblies, fasteners, and possibly some parts are put together into a higher level subassembly. This pattern proceeds until the final assembly takes place, and the finished product is ready. Alternatively, this pattern can be visualized by beginning with a complete product – say an automobile – and breaking it down progressively into its components: first, the body, the chassis, the engine, the drive train, and the bolts and nuts which hold all these together. Then the body consists of the outer shell, four doors, the hood, and some hinge pins. The hood consists of the outer layer, the inner layer, the latch mechanisms, and the hinge arms, and so on down to the spring that holds the release lever on the latch.

The structure of the assembly operations is obviously a series of subassembly operations in which two or more parts are positioned correctly relative to one another, and fastened there. The fastening may be permanent, as when the inner and outer layers of the car hood are welded together, or may be subject to disassembly, as when the cylinder head is bolted to the engine block. Some assemblies hold their components in fixed relationship, as with the cylinder head, while other assemblies hold their components in moveable relationship, as when the hinge pins are put in to hang the doors to the body. The variety of assembly configurations, and of fasteners, is almost infinite.

One tends to think of assembly functions as following after part fabrication operations, but there are occasions when two parts are assembled and then further machined because the final machining operations cannot be performed on the parts separately. Such partial assemblies become, for the purpose of planning and scheduling, a single item forming the input to the later machining operations.

Subassembly operations are planned by following the structure of the bill of materials. A list of all the parts of the subassembly, sometimes called a "kit list," is written. It forms the basis for picking from the store of completed parts the necessary number of all the needed parts and fastenings. In many instances subassemblies are made in batches of some convenient size, and when their physical size is appropriate, are put together in a bench operation. Completed subassemblies then acquire an identifying number and become components of the next level kit list.

It is convenient to list the kit in the order in which the worker would logically pick up the pieces. Each of these steps may be assigned an elapsed time allowance, and they, in turn, may be totaled for the subassembly. This figure is used by the schedulers in planning the date and time when each unit must be completed. As in fabricating parts, a certain amount of time must be allowed as a cushion to take care of problems such as parts that do not quite fit, missing parts, damages during assembly, and the like.

Tests or inspections during assembly may be required, and if so they are treated just as if they were subunits of the assembly sequence. There is obviously no point in continuing to fasten things together if the assembly will not pass the ultimate inspection and test. Equally obviously, if the inspection rejects a subassembly, it must be reworked. Either it must be disassembled and the aberrant part replaced, or the parts properly repositioned and refastened. This will take time, and may require replacement parts from stock. If no stock parts exist, then everything must be held until a new part is made. All of this will enter into the scheduling and budgeting function, and the planner will wisely build some "cushion" time into the overall schedule to cover just such events.

## OUTPUT FROM PRODUCTION PLANNING

Refer to Figure 9.1 to visualize the output from Production Planning. It consists chiefly in instructions to the shop—Box 6 in Figure 9.1—and in directions to resource and materials procurement – Boxes 4 and 5. Figure 9.2 shows another pair of outputs fed back to up-stream functions: the final bill of materials and production data, as well as the special tooling and scheduling requests, go back to Box 2, labeled "Make and Administer Schedules and Budgets." Also problems which require changes in product designs are sent back through channels to the Development function.

Production instructions go to the shop. This includes for each part a work packet containing the work order, the operations list or routing, the work instruction sheets, the NC tapes, and any necessary drawings, all as described above. It also includes for each assembly and test operation the assembly phasing lists to guide the progressive assembly, and the test specifications for the ultimate product performance verification.

Tooling specifications for each operation on each part, and for each step in assembly, are sent to the unit which provides production resources, Box 4 of Figure 9.1. Here the word "tooling" is used not for the machine tool or its equivalent, but for the nonexpendable adjuncts necessary for the specific operation. This might include holding fixtures or jigs, made to cradle and hold a specific part on a specific machine tool at one specific stage in its fabrication. When such tooling already exists, as in repeat orders for fabrication of some part, the tooling will merely be retrieved from storage. When tooling is needed and none exists, it is necessary to either build it in-house or purchase it from a specialty vendor. If it is to be built in-house, the request goes to a design function analogous to the design function that designed the product, but conventionally located in the Production domain. The tooling is designed, its components detailed, and it is produced in a Tool Shop in just the same manner as are the parts comprising the product itself, except that tooling is usually one-of-a-kind fabrication instead of multiple fabrication. Tooling is stored when it has served its initial purpose, in anticipation of later reuse.

Similarly, tooling may be required for assembly operations. For example, in building aircraft, very sizeable assembly jigs are built to hold the wing or the fuselage components in proper relationship while drilling and riveting is done. These serve in the assembly of every unit of that particular aircraft model.

Production Planning sends to the unit (Box 5) which procures materials the list of all materials needed, and the quantities of each, which will be incorporated in the final product. Production Resources, Box 4, adds its list of materials needed to make the tooling, and the quantities. Procurement may find adequate supplies in stock, or may have to order the materials from the various vendors. Some of the procured materials may be sent back to Production Resources to be made into tooling or other adjuncts not destined for incorporation in the product. The bulk of the materials and purchased components are sent to the shop for fabrication and assembly into the product.

Production Planning feeds back to the schedulers and budgeters all the myriad details of this intricate exfoliation of planning. The manufacturing bill of materials forms the key to the whole structure of scheduling and budgeting at a detailed level. All the operation times are reported, to be converted both into costs for the budgets and into time spans for schedules. Materials specifications similarly can be converted into estimated costs and delivery times for procured materials. Set-back times can be calculated from the required delivery date to determine the assembly schedules, and from them the set-back times can be calculated for the procurement of materials and the start of parts fabrication.

Discrepancies between the assumed schedules and budgets, made before Production Planning did its work, and the final set of schedules and budgets may require correction to some part of the plans, or an appeal to the Project Manager for administrative action.

Production Planning receives back from the shop, and from the resources and materials procurement units, any and all sorts of problems. As seen in Figure 9.2, Resources may reflect difficulty in supplying on schedule the required equipment, tooling, skills, or the necessary technology. Materials may report the unavailability on schedule of specified materials or procured components, or the excessive cost of some procured item relative to the cost assumed in budgeting. The shop may report all manner of problems causing production to fall behind the schedules or budgets, such as absence of skilled workers, or strikes; power failures; equipment breakdown or malfunction; materials off-specs; quality control rejections; and many other unforeseen problems. All of these problems call for correction of the plans to conform to reality.

If it appears that a change in the product design is needed, this request must be sent back through channels for decision and appropriate action. This feedback is shown as the second output (02) from Box 3 in Figures 6.5 and 9.2. Design changes may be requested because of difficulties encountered in assembly and disassembly, problems due to tolerance build ups, part failures in product testing suggesting design calculation errors, ideas for cost reduction or time-saving perceived by the shop personnel, and many other reasons.

## SUMMARY

Planning for Production is the third of the six functions which comprise Production. It converts the product design, as received from Development in the form of layouts, detailed drawings, and a bill of materials, into detailed manufacturing methods and instructions which conform to the Plan for Manufacturing and the Schedules and Budgets.

For each element of the design (part) a make or buy decision is made, and a list of parts to be manufactured is prepared. For each part to be manufactured, a sequence of operations is prepared. For each operation on each part, instructions are prepared showing the equipment to be used, tooling, skills, and time allowances. For each assembly sequence, an assembly phasing list is prepared, showing the sequence of operations and fixtures required. For components to be purchased, procurement specifications are written.

The production planners send the fabrication and assembly instructions to the shop; the materials specifications and quantities to the procurement unit; and the tooling and equipment requirements to the resources unit. The myriad bits of data generated by this planning process provide the basis for cost compilation for budgeting, and the timing setbacks for scheduling. Discrepancies between the assumed schedules and budgets, made before Production Planning did its work, and the final set of schedules and budgets may require correction to some part of the plans, or an appeal to the Project Manager for administrative action.

Difficulties in actually meeting the plans are reported back to Production Planning for modification of their plans to work around the problems, or possibly to request a change in the design of the product to alleviate the difficulties.

# Chapter 10

# Production: Provision of Resources

The provision of production resources is the fourth of the six functions comprising the Production function. It is shown in Box 4 of Figure 6.5, labeled "Provide Production Resources." It is the function responsible for providing facilities, equipment, tooling, people, and technology to meet the needs set forth in the plan for production. These resources are distinguished from the procurable items (Box 5) by the fact that they are not incorporated in the final product, but still are required to produce that product.

The provision of each of these five kinds of resources may proceed substantially independently of each other, excepting that each must be available when needed to meet the production schedule. It is therefore necessary that plans for each item to be procured be made as early as possible in the course of events. However, while the procurements may be independent, the five categories are interdependent. Facilities and equipment are related to one another through the space required by machinery and similar equipment. If new capital equipment must be added to the factory, there must be space to house it, or more buildings must be obtained. Equipment and tooling are related for obvious reasons: the tooling must fit upon the machine tools on which it will be employed. Worker skills are related to both machinery and tooling; any one of these three may place constraints upon the other two. Finally, technology enhancement may well call for changes in all four of the other categories of resources. All five must meet budget and schedule constraints.

This may be a good place to define the many meanings of the word "tool." A "machine tool" is a powered mechanism, fixed in place in the factory, for performance of some particular function. Example: a lathe or a drill press or a forge. "Tooling" is the term applied to nonexpendable items used in conjunction with a machine tool for the production of some specific part. Example: a jig, a fixture, a template, or a forgoing die set. "Tools" or "Cutting tools" are general purpose but expendable items which may be used to shape any part, but will need sharpening or replacement from time to time. Example: drills, milling cutters, shears, reamers, and the like. All three of these uses describe a kind of manufacturing resource.

## PROVISION OF FACILITIES

The provision of facilities encompasses the planning, building, modification, operation, and maintenance of the various structures that will house the production operation. In an on-going enterprise, facilities additions will be made in response to the equipment needs spelled out in the resource specifications received from Production Planning. When new machine tools will occupy more space than is available in the factory, new buildings must be acquired, or the operations that require the new machines must be farmed out to a subcontractor.

Building a new facility is a relatively infrequent task, and therefore involves the aid of outside specialists – architects, real estate people, bankers, and building contractors. It is not undertaken lightly, because of the large one-time expenditure, nor is it a short-time activity. Long range planning is, therefore, important to ensure that the plant will be ready when the new machinery is delivered, and that it is ready within budget. On the other hand, building modifications may be required frequently. New access doors or air conditioning equipment may be installed. Machine tools may be rearranged on the shop floor to give more advantageous flow paths for work moving from machine to machine. One company undertook the construction of large nuclear reactor vessels that required the use of a huge vertical turret lathe; to get the work pieces on the lathe table, the columns had to be raised 10 ft., which meant that the roof in that bay had to be raised!

An alternate to building new factory space, or modifying existing factory buildings, is the renting of some conveniently located and available factory building. This will invariably call for the reworking of the structure to some extent before installing new machinery, and the provision of interplant material movement, personnel facilities, etc.

Operation and maintenance of facilities include a host of mundane but essential activities that one rarely thinks of when considering manufacturing as a whole. Facilities must be heated in the winter, cooled in the summer, ventilated, cleaned, painted, plumbed, lighted, and the exterior cared for by gardening or lawns if space provides such amenities. Parking lots for employee's cars, water, power, and telephone services must be provided. Janitors are required to keep the plant clean and watchmen provided to assure security. Waste materials must be disposed of in conformance with OSHA and local regulations. Employee food services and recreational facilities are usually required. No one of these activities is very glamorous, but they are nevertheless essential. They fall to the facility group.

## PROVISION OF EQUIPMENT

Provision of equipment encompasses the development of equipment plans, the acquisition and installation of the equipment, control over the equipment, and its maintenance by both routine and emergency work. The suite of machine tools is structured to meet the plans for manufacturing all of the products in manufacture at any time. When a new product is introduced for which adequate machinery is not in place, new tools must be bought or old tools modified. The alternative is, of course, to farm out that part of the production work to a subcontractor.

Equipment purchases are, like building acquisitions, long lead-time operations involving relatively large amounts of money. They are therefore carefully planned well in advance of the time when they are needed. Purchase of new equipment is usually justified by the profit center which will use the tool, on the basis of the savings its use will effect over the use of current tools. (Acquisition of new systems or new technology is, on the other hand, usually justified on a company wide basis because the savings expected will come from economies in many or all of the profit centers.) Justification of new capital equipment is a very complex subject, beyond the scope of this book, but one that must not be overlooked.

Equipment may be bought, or it may be leased. The decision is a matter of economics easily settled by the Corporate Treasurer. On the other hand, equipment may sometimes be made advantageously in-house. There may be some machine tool which is available and which can be modified by the company tool room to perform to meet the new requirements. This has the advantage that the tool may be designed precisely to meet the requirements, rather than acceptance of the capabilities of the nearest commercial tool. Also the tool to be modified may be old enough so that its original cost has been largely written off, and the new tool will therefore represent a smaller capital commitment. There are many such tools in industry that are dedicated to one single production task without undue cost when they stand idle from time to time.

When a major machine tool is purchased, the acquisition is done on the basis of carefully and explicitly worded specifications. A trial run of real parts is made in the machine vendor's plant before shipment is approved, and again after installation in the company's plant. It is usually the responsibility of the buyer to prepare the machine site and foundation, if one is required. Operator training may be provided at the vendor's plant so that the machine can go into use as soon as it is installed and has passed inspection. Once accepted, the Resources unit is accountable and responsible for the machine.

Equipment is not invulnerable to wear and damage; it must be inspected regularly, and if necessary it must be readjusted, recalibrated, or even rebuilt. Modern machine tools lose about 50% of their productivity in the first 10 years of their lives, and at the same time suffer from technological obsolescence when compared to newer tools. Rebuilding can restore some of this loss. Accidental damage or failure of equipment must be corrected on a priority basis; it is one of that large class of unplanned nontrivial problems which plague the schedulers.

Equipment must also be routinely maintained – cleaned, lubricated, and realigned. Some of this is the responsibility of the operator, some of it devolves upon experts in the Resources facility. It is easy to understand that the provision of equipment is another unglamorous but extremely essential function in any manufacturing enterprise.

## PROVISION OF TOOLING

The provision of tooling encompasses the supply of all the items used in the production of a specific product. As explained earlier in this chapter, tools and tooling are used in conjunction with a machine tool in the course of production. (The machine tools were discussed in the previous section of this chapter under the more general term "equipment.") Tooling may be expendable or nonexpendable. This means that some tooling may be used and reused as long as the workpiece for which it was designed is being produced; aside from minor maintenance, it is unchanged by use. Other tooling may wear or deteriorate in use, and may require periodic resharpening; after a few such regrindings, it becomes irrepairable and a new unit must be procured.

Some tooling may be unique to one single workpiece, as for example a clamp designed to cradle and hold some peculiarly shaped part; other tooling is of standard design, as, for example, twist drills or end mills that are economically available in a variety of standard sizes from commercial tool supply houses. Tooling may be made in-house or purchased from subcontractors. However expendable or nonexpendable, unique or standard, home made or purchased, the supply of tooling requires careful management.

The tooling to be used at each operation on each part, unless very obvious, will be specified on the operating instructions. Unique tools will be identified by a number for identification, and as an aid in retrieval after storage if they are used intermittently. Thus when the production schedules show that a batch of parts is moving to the next operation on the route sheet, the materials handling agency will know that these unique tools must be taken from storage and moved to the machine tool site. The same applies to the standard tooling, except that it is identified by its dimensions rather than a tooling number. Any one of the 0.995/1.000 in. standard HSS twist drills in stock should be as good as another.

Tooling is usually thought of in several classes. There are passive tools which serve to hold a part on a machine tool. This includes chucks, clamps, vises, as well as jigs and fixtures. [A fixture is a special holding tool for some part at some stage of its production. A jig is the term applied to a fixture with a built in set of guides to steer the cutting tools (drills, e.g.) to the precise location where a hole is wanted.] There are tools which participate actively in the shaping of a part by a machine tool, such as cutters, bending forms, etc. There are measuring tools which neither hold nor shape a part, but check the performance of previous operations against requirements. This class includes standard and special gauges, micrometers, etc. Finally there are hand tools: hammers, wrenches, scribers, brushes, and others in endless variety. Each of these classes of tooling has its own supply industry and vendors.

Tooling of all sorts which is to be made in-house goes through a design process similar to that used in the design of the product itself. The several steps described in Chapter 5 apply in abbreviated form to the development of tooling or special machine tools. For very low production quantities it may be economical to support a work piece in a cradle made of modular reusable blocks and clamps. Tooling design follows a far more stereotyped format than does the development of new products because the tooling differs from instance to instance principally in dimension and shape, and very little in function. Furthermore, tooling is a one-of-a-kind design problem; there are few concerns about multiple manufacturing techniques. Testing the tooling design is accomplished when it goes into use, and any corrections needed are immediately reported, and immediately effected. While the functional structure of tool design is the same as that of product design, it is customarily conducted in a separate, dedicated design facility.

Tool production is also functionally identical to the production of the principal product but is customarily conducted in a shop called a "tool room," equipped with precision machine tools adapted to one-of-a-kind production. The need for precision is important. End product parts are made to dimensions each of which is stated with a tolerance. The tolerance tells how much random variation is to be tolerated in the actual size of the part. For example, the spacing between the bearings for a pair of shafts connected by a pair of gears may be dimensioned 8.250/8.256 in., meaning that any actual distance falling between those two distances will be acceptable. The part tolerance is $+/-0.003$ in. Now when the tool designer dimensions the jig which will guide the drilling and reaming of these two holes, he will allow one tenth as much tolerance: 8.2527/8.2533, which says that the jig must be constructed within three tenths of a thousandth of an inch. This requires a superior type of machine tool. Such added precision also adds cost – roughly twice as much for each added order of magnitude in the tolerance figure.

Expendable tools must be sharpened or otherwise restored as they wear during use. Tool sharpening is a highly skilled art requiring special machine tools designed for the purpose, and is usually conducted by a specialist who not only sharpens the tools but checks the end result of his work and then stores the tools until they are again needed. From time to time a tool wears beyond the point where it may be restored, and it must then be scrapped and replaced. Some cutting tools used in numerically controlled machine tools must be mounted in tool holders which are in turn held in the machine tool spindle. The cutting tip of the tool must be positioned accurately at a prescribed distance from the reference datum on the tool holder, so that when the automatically controlled tool makes a cut on a work piece, the cut surface will be where the manufacturing engineer planned for it to be. Such tools are called "preset" or "qualified tooling." The presetting, and its checking with an optical micrometer, is done by the specialist in the tool room.

All the tooling discussed in the foregoing must be accounted for, stored when not in use, protected against rust, issued when the production plans call for it, and then retrieved, reconditioned, and restored.

## PROVISION OF PERSONNEL

The provision of personnel to operate the equipment and to manage the sequence of operations and the flow of work through the manufacturing process is an essential function. In many instances the provision of trained people is headed up in the Support function of the enterprise, rather than in the Manufacturing function. Figure 2.1 shows this support function in Box 4, labeled "Support Enterprise." It is also shown more specifically in Figure 2.2, the node tree diagram, as "Manage Personnel."

The management of the personnel function requires skills quite different from those involved in the management of either manufacturing or of marketing, and is quite properly left in the hands of the experts. It is a subject outside the scope of this book, but a few observations on the subject of personnel are appropriate.

In assembling the cost elements of a product, particularly those costs accrued through the application of worker's time, it has been customary to make a distinction between direct and indirect labor, and between direct and overhead labor. The handbooks define direct labor as that which changes the form or advances the stage of manufacture or assembly of a product; its cost may be allocated directly to a specific product. Indirect labor costs are sums paid to direct employees for overtime or night work premiums, vacations, insurance, and the many other fringe benefits; it may be treated as a mark-up on the direct labor costs. Direct plus indirect labor costs represent the cost of retaining and applying the human effort necessary to fabricate, assemble, and test the products.

Overhead labor costs represent those human efforts necessary to maintain the facility and equipment in working order; to supervise the direct labor; and to plan, schedule, budget, and control the flow of work through the plant. It is not directly allocatable to any specific part or product, but is essential to the conduct of the manufacturing operations. Its cost is therefore spread over all of the work in progress in some equitable manner.

For example, the operator of an NC machining center would be a direct worker; his time would be directly related to the parts he produced in a day. However the engineer who wrote the NC program used in the machining center would be an overhead worker, because his programming would be done but once for any given part design, and the control tape would be used every time that part had to be produced, be it once or many times. The maintenance crew who diagnose malfunctions and repair equipment are essential insurance for the continuous availability of the equipment; ideally, they would never have to do anything, but in reality they serve whenever and however they are needed to keep production going forward. They, too, are overhead workers.

With the advancement of manufacturing technology, and particularly with the adoption of automated machinery and computer integrated manufacturing technologies, the ratio of direct to overhead workers is changing, and the number of workers in the two classes is changing. There are fewer direct workers, and more overhead workers per unit of output. Automated machinery, once it is set up and working properly, will continue to produce with very little human assistance. However to get it set up correctly requires special skills not needed for manually controlled machinery.

The educational level and the operational skills required are widely varied in any given organization, and with the advent of higher technologies, the spread between the higher skills and the lesser skills becomes greater. This change is principally due to the raising of the level of the upper end of the spectrum to cope with the advancing science of manufacturing. As workers are trained up to these new levels, or as new people are hired to fill these needs, the ratio between high and low skill levels increases. However most factories have a mix of all skill levels.

As automated machines are introduced, the control of their functions is transferred from the human operator to the internal automatic mechanisms. For example, the operator of a standard horizontal boring mill has to be able to place a work piece on the table, align it accurately, clamp it, select the correct cutting tool and place it in the spindle, select the correct spindle speed, and move the tool into the work the correct distance which he reads from the machine scales or his micrometer. However when this machine is converted to numerical control, with automatic tool and pallet changers equipped with fixtures, his role changes. He sets the part in the fixture and closes the clamps; he starts the control tape; when the machine is done with the part, it stops and ejects the pallet so he can take off the finished part. Note what has happened. The NC programmer has decided all the speeds and feeds for all the cuts. The tool designer has made the fixture so that the part is correctly positioned. The NC machine changed its own tools and made its own measurements. In effect the operator is now just a loader/unloader and a monitor, while the principal skills of the direct operator have been transferred to specialists elsewhere. What is more likely, the skilled operator has now been trained and promoted to be a programmer or a tool designer, and a new employee with much lower skill levels has been hired as the monitor. Parenthetically, at the same time the above change was taking place, the time per unit of production dropped, and became far more predictable as the human element moved farther from the controls of the machine.

Not only are operator skills changing, but managerial skills are changing with introduction of data processing equipment to the manufacturing scene. Shorn of job specifics, a managerial task consists of three parts: receipt of information, making decisions, and order/report issuance. These three steps have always been the rule. Data processing equipment offers the manager instant access to the latest status reports at the touch of the keyboard. It offers him the ability to call upon the computer to apply a decision making algorithm if there is an appropriate one, or to run a simulation program so he can select what appears to be the best decision. Additionally he can transmit his decision to his subordinates in the same direct manner. His function is still the same, but the tools at his disposal to perform his function are new, and if he is to use them well, he must learn a whole new group of skills.

Finally, safety of personnel is an important responsibility of management. It is directly related to the processes being used in a plant, to the machinery, to the materials, and to the supervision. The provision of all these resources must take into consideration the assurance of the highest achievable safety in the personnel's environment.


## PROVISION OF PRODUCTION TECHNOLOGY IMPROVEMENTS

The provision of improvements in a company's production technology assumes that a production technology is in place and operating, but recognizes that it can always be improved. The search for enhancement goes on continuously. Here again, the rapid evolution of manufacturing technology makes this function a matter of active concern for those charged with the provision of resources. It is a key element in Quality Assurance.

Perception of opportunities may take place almost anywhere in the organization. People who have contact with other similar industries, or who attend professional society meetings where such matters are discussed, may see opportunities. Contacts in the academic world may open vistas of new technologies in their developmental stages. Vendors who call upon the company are frequently sources of ideas, because they also call upon other similar users of their products and observe their methods. However it is usually not adequate to leave technological progress to happenstance. Many enterprises have specially focused research committees to address this subject.

The perception of a possible technological enhancement comes most successfully from a specifically assigned study group on a corporate level, rather than from a profit center within the organization. The reason for this is that a change in the established way of doing things is generally resisted, and the more radical the change, the more stubborn the resistance. A mandate from the top level of the enterprise is sometimes necessary to bring things to a head. Furthermore, major changes in technology will inevitably affect many parts of the company, and it takes a high level thrust to overcome parochial interests. Finally, the more innovative the enhancement, the greater the chance that there will be some considerable doubt as to its successful introduction; its adoption will be a gamble. The only person entitled to take such big gambles is the Chief Executive Officer, the man at the very highest level.

Economic justification for minor improvements usually can be established in the unit most benefited by the change. For example, the purchase of a new machine tool using new technology will usually be justified on the basis of its replacement of some existing tool with a resulting reduction in cost, or time, or labor, or improvement in quality. The profit center responsible can assume the responsibility. However if major changes are to be made, then several profit centers may be affected, some of them adversely and some of them profitably. Only the overall interest of the whole company should govern in such cases, but it may take an appeal to a higher level manager to settle internal differences.

Economic justification for major system innovations will require a very high level decision. For example, the acceptance of the concept of computer integrated manufacturing will impact every single corner of a company. It will change managerial techniques, product designs, marketing opportunities, and manufacturing facilities. It will affect skills required from the board room right down to the shop floor. It will also require a very sizable investment. Furthermore, it is not easy to predict with certainty the return on the investment. Embarking on such a major change is an irreversible decision. The only person in a corporation authorized to take such a gamble is the Chief Executive Officer.

Large or small, the introduction of new technology will take three stages: the acquisition decision, the adaptation to the company environment, and the implementation. The more sweeping the change, the more important it is to plan the overall change, schedule it carefully, and then implement it piece by piece.

## RESOURCE PROBLEMS

As may be seen, the provision of resources is a group of diverse activities, each essential to the overall conduct of the production function. It is essential to remember that for every operation on every part, and for every assembly operation, there must be present at the site and on schedule:

A workable machine tool;
The work in process;
A competent operator;
The necessary tooling—cutters, fixtures, gauges; and
The packet of instructions.

Problems are inevitably encountered in the acquisition processes, and delays will have an impact on all the rest of the production unit. The Resource function is responsible for supplying most of these items. Accordingly, resource problems are reported back to the production planners, and to the scheduling and budgeting group. This feedback of information is shown in Figure 9.2.

Further reference to Figure 9.2 will show that the Resource function will also report resource availability status to the scheduling and budgeting group, Box 2, and information on the manufacturing capability of the plant to all concerned. They will also receive from time to time from the shop, Box 6, requests for delivery to the shop of some needed machine, tool, fixture, or service.

## SUMMARY

Provision of Production Resources is the fourth of the six functions which comprise Production. It provides facilities, equipment, tooling, people, and technology to perform the production functions, but which are not incorporated in the final product. Furthermore, the resources must be available on site at the times when they are needed to aid in each operation on each part of the product, and at each assembly operation.

Provision of plants and machine tools takes time and large amounts of money, hence must be planned well in advance of their moment of need. Tooling may be built in-house or purchased, but has a shorter lead time requirement. Staffing the plant with the necessary people skills is a continuing problem, increased when new technology is to be introduced. The provision of all five classes of resources are interrelated tasks.

Plants, tools, and machinery must be maintained, stored, repaired when necessary, and accounted for.

The introduction of advanced technologies has an impact on every facet of production, and hence is best initiated at high levels in the organization. New technologies also change the mix of personnel skills required, with a shift toward higher skill levels and greater use of "overhead" personnel.

# Chapter 11

# Production: Obtain Production Materials

The procurement of production materials is the fifth of the six functional parts of Production. It is shown in Box 5 of Figure 6.5. Its function is to procure all those materials and items which will be included in the delivered product, or which will be converted into resources used in the production process.

If materials are to be converted into finished products by the manufacturing process, then those materials must be procured. This includes a wide variety of things:

Steel bars and slabs, aluminum sheets, plastic monomers, wood products, and the like;

Finished components such as motors, bearings, hardware;

Parts made by subcontractors to the user's designs;

Cutting and forming tools, gauges, fixtures, etc., made to order for the user;

Standard cutting tools such as twist drills or milling cutters;

Hand tools—hammers, wrenches, screw drivers, etc.: and

Supplies such as paint, oil, and office supplies.

In addition, this function may be the agency that places the contracts for new equipment in response to directions from the Resources unit.

The function is frequently known as "Purchasing" or "Procurement." The organizational position of the Purchasing, Receiving, and Warehousing functions varies widely in industry.

This function not only procures all the needed items for manufacturing, but it maintains an appropriate inventory from which it can furnish quantities to the shop upon requisition, in accordance with the production schedules. Reference to Figure 9.1 shows several controls which govern the procurement function. Box 1 in Figure 9.1 provides the overall plan for manufacturing of each product or project. This acts as an advance warning of long lead time items whose procurement may have to be started long before the item is needed. It also gives the general plan of the requirements for the project. Box 2 in that figure provides the schedules and budgets for the procurement, yielding a time table for actually being prepared to deliver the items to the shop. Box 3 provides the details of materials and quantities for each and every item to become part of the product, and Box 4 provides the procurement list for things which will become parts of the manufacturing resources.

These last elements are shown in Figure 9.2 as outputs from Box 5, transmitted to the Resources unit, Box 4. If they are standard items they simply go into the appropriate store room, but it, for example, it is steel to be processed in-house into fixtures, it goes to the shop's Tool Room for manufacture, thence back to the proper store room. However, having been converted into a fixture, it then falls under the jurisdiction of Resources.

Funds for the procurement of materials originate in the funds allocated to Manufacturing by the enterprise management as shown in Figure 3.2, as the input to Box 1 of that figure. They are not shown explicitly in any subsequent diagrams, but are implicitly transferred to Production in the directive to produce, shown as a control to Box 3 of that figure. Here we also see the input of procurable items which will be the source for the procured items we are discussing. Figure 6.5 is the decomposition of the Production function, showing in detail the transmission of the Directive to Produce, C2, from Planning for Manufacturing to Materials Procurement, Box 5, where it meets the available Procurable Items input.

The principal flow of materials through this function is shown in the figure as the input of procurable items, and the output of materials and components to the shop. What the function does to these inputs is to receive, inspect, accept, store, and issue to the shop.

Ordinarily the materials would be sent to the shop in accordance with the production schedules, but there may be occasions when the shop needs more than it received in the normal course of events. A piece of material may be spoiled in production and rejected by the inspectors, and a new piece has to be started through the works to replace it. A piece of material may prove to be defective in the course of production, as for example the discovery of a blow hole in a casting at some critical point. In such cases, and for other valid reasons the shop may call for additional materials; these requests are shown in Figure 9.2 as another control on the operation of Box 5. Similar requisitions may also be sent by Resources, Box 4.

Besides the actual flow of materials, Figure 9.2 shows two other outputs from Box 5. The status of the inventory is sent back to the administrative unit, Box 2, so that they are constantly informed of the supply situation. Problems with supply, such as slow delivery or unexpectedly high costs, are also sent back to the scheduling and budgeting people in the administrative function, Box 2. Such information may in turn trigger changes in the Master Schedule, or extra efforts to correct the difficulty without upsetting the schedule.

Procurement determines the total materials needs of the enterprise as a function of time, as distinct from the needs of one particular product or project. It is probable that the production plans of many different products will call for the provision of the same material item. Procurement consolidates all these various needs, listed by quantities and the dates when they will be needed. It then examines the existing inventory of those materials, and makes an allocation to the individual projects. When it appears that there will not be enough material in inventory at some specific date, it initiates acquisition of the necessary amount of the material.

It is the objective of the Materials function to minimize the inventory of materials and procured items insofar as possible without risking any delay in the production schedules, of which they have been appraised. The reason is that inventory is expensive; there is the interest charge against the money tied up in inventory and not otherwise profitably employed. There is the cost of providing, maintaining, and operating the storage facilities. There may be deterioration of materials held in inventory or pilferage from the warehouse stocks; and there is always the chance that the need for materials held for future use may evaporate due to changes in designs or in product demand. Therefore, it is usually the case that Purchasing will seek to order as little material as possible, and still be assured that it will be available on the date for its scheduled use.

While material acquisition is planned to provide only the necessary amounts on schedule, purchases may be made in larger quantities to take advantage of the economies of bulk purchases. The savings in unit cost may offset the cost of carrying the inventory, or there may be uncertainties in the future availability of the item.

When the enterprise is engaged in the manufacture of a series of batches of some product, parts are usually made in lots determined to be the minimum economic lot size to put through the factory. Owing to the cost of making setups, which may well exceed the machining cost on a single unit, it is desirable to make a number of parts in a batch, and this economic order quantity may be greater than a batch size. Surplus parts are stored until the next batches are ordered, or to serve as a reserve for spare parts orders. The advent of computer integrated manufacturing and flexible manufacturing systems is reducing the minimum EOQ, and ultimately it may approach one. But until it does reach one, parts are usually made in batches.

Group Technology may suggest that several batches of parts can be profitably grouped together and manufactured as a batch, saving the cost of several setups and varying only the machining dimensions. This also tends to increase inventories, but is only used when the total cost of the procedure is a reduction over the total cost of producing the parts separately.

Finally it may be that in slack production periods, Management may decide to work ahead, building inventory of some parts, thus leveling the work load on the factory over a long period of time, and retaining skilled employees rather than having to lay them off.

It is the responsibility of Materials Procurement to take cognizance of all these policies and to provide the necessary input materials. When an acquisition plan has been made, requisitions are issued to the Purchasing group.

Purchasing then selects the source from which the material is to be purchased. The Purchasing Agent will have a list of vendors of each of the many classes of items normally bought, and will select a vendor from the list. He may get quotations from several, or may make his selection on the basis of past performance of some one vendor. If the enterprise buys large amounts of a material relative to the total business of a vendor, the Purchasing Agent may elect to spread large purchases over several vendors. Buyer and vendor then agree on price and delivering date, and the specifications by which the delivered material will be inspected for acceptance.

When the order has been placed, a follow-up file is opened to monitor the delivery of the items, and the acceptance criteria are sent to the Receiving Department. From time to time a vendor will fail to deliver on time, and appropriate action must be taken. Alternate sources may have to be found or production schedules changed. The administrative unit is kept informed.

Upon receipt of the material, it is inspected in accordance with the criteria set up at the time of purchase. Quantity will be checked, and more or less extensive test procedures may be performed, depending on what the material is. Rejection of some of the material will be cause for negotiations with the vendor and the assurance of adequate replacement material. When the delivery is completed, the order is closed and Accounting notified to make payment.

Accepted material is placed in inventory. This is not only an accounting category to deal with the investment in the material, but also an actual warehouse storage. Storage and retrieval of materials are problems of location and of proper storage facilities. Large slabs of steel or big castings may be stored out of doors or on covered floor spaces. Tiny electronic components may be stored in boxes on shelves or in some similar safe place. Automated storage and retrieval facilities are available for all sorts of materials; they provide a means to store things safely, and then get them out when needed with a minimum of labor and a minimum of loss. Commonly used hardware items, such as nuts and bolts, may be stored in bins on the assembly floor from whence they may be taken by the workers as needed without a formal requisition.

Some items kept in storage may present special problems in warehouse construction and operation. This includes materials which are

Toxic, such as many chemicals and certain drugs;
Light sensitive, such as photographic emulsions;
Pilferable, such as fire arms and ammunition, radios, TV sets, etc.
Refrigerated items, such as food substances and some drugs;
Very expensive, such as jewelry, precious metals, etc.;
Very delicate, such as instruments.

Appropriate warehousing and security must be provided in these and similar circumstances.

Because parts that are purchased as finished items, such as hardware and electrical components, are usually bought in quantity, the warehouse will upon demand go to the storage location and withdraw the required quantity from the supply, adjust the inventory records, and deliver the items to the requestor. Similarly, steel bars and rods, wire, and the like are also bought in quantities. However, when a request comes through for ten pieces of ½ in. round steel bar of some certain type and 8 in. long, the warehouse must cut off the pieces from the stocked material, and issue the pan full of parts. While this might be regarded as the first step in fabrication, and hence properly a function of the shop, it is commonly considered to be a function of Procurement.

Castings and forgings are frequently purchased from vendors in these specialized fields. While the operations of casting or forging are properly fabrication functions, and would appear on an operations list of a normal complete set of production instructions, purchased castings and forgings are frequently regarded as procured items. They are ordered and stocked until the schedules call for machining them, and then they are issued to Fabrication for the production of some specific part.

## SUMMARY

Obtaining the Materials for Production is the fifth of the functions which comprise Production. It procures from vendors the materials which will be converted into the end-product, as well as the materials which will be made into production resources. These materials may be raw material or finished items. They may be standard items, or made to order for the enterprise.

Materials in inventory are matched to requirements for production, and the need for purchase of more material is determined as a function of time. Sources are selected, orders placed, and followed up. Upon delivery, acceptance of the material is contingent upon its passing inspection. Accepted material is placed in inventory and issued to Production upon requisition or in accordance with schedules.

# Chapter 12

# Production: Conversion of Materials into Products

Production of the product is the function that alters the form and character of the input materials into the finished component parts, which then are progressively assembled into the end-product. It is the sixth and last of the six parts of the Production Function. It is shown in Box 6 of Figures 6.5, 9.1, and 9.2. It receives the purchased raw materials and purchased finished components from the Procurement Function, Box 5, as well as production resources from the Resources Function, Box 4. Using the resources, it converts the raw materials into finished parts in accordance with the production instructions received from the Production Planning Function, Box 3, and assembles the parts and the procured components into finished products. When the completed products have satisfactorily passed the acceptance tests, they are delivered to the Marketing Function for distribution to users, as shown in Figure 2.1.

This fabrication, assembly, and testing function is complex, and varies widely in its complexity depending upon the product being made and the user industry being served. We discussed the diversity of manufacturing in Chapter 1. Before delving into these details, however, some basic comments are in order relating to the function as practiced in any industry, regardless of the product.

Production of parts and products is controlled by several inputs. The overall strategy for production will be determined by the Manufacturing Planning Function, Box 1 of Figure 9.1, as discussed in Chapter 7. The major structural divisions of the product will naturally determine the assembly sequence. The make or buy decisions on components will determine the amount of work to be done in the shop. The manufacturing technologies selected will determine what new facilities – plant buildings, machinery, tooling, and operator skills – if any, must be acquired. The overall schedules and budgets set up will determine priorities in shop scheduling. Quality requirements will determine testing procedures, and may even cause the Quality Assurance people to suggest modifications in production methods.

The detailed strategy for production of each part of the product will be determined by the Production Planning Function, Box 3, as discussed in Chapter 9. The bill of materials is the key to the massive amount of details in the production plans. For every part there will be a sequence of operations list, and for each operation there will be a listing of the machine tools, the tooling, the operator skill, and the instruction sheets required. For each step of the assembly there will be an assembly phasing list showing the sequence of assembly. For each of these many activities there will be an estimate of time required and the cost which is expected to be generated – information which will have been fed back to the Scheduling and Budgeting Function. All of this production planning will be fed to the shop and act as a control on their activities.

Scheduling and Budgeting, Box 2, will also act as a control upon the shop. They will begin with the broad brush estimates of Manufacturing Planning, and refine their data in accorance with the detailed cost and time estimates generated in Production Planning. Furthermore, they will receive from the shop the actual times and costs expended, as work progresses. These actuals will be compared to the initial plans to monitor the conformance of the work to the targets set by higher authorities.

In addition to the cost and time reports rendered by the shop, requests may be sent to the resources people when additional equipment, tooling, or services are required, and to the procurement people when additional materials are needed. When problems arise in the shop which threaten either the established budgets, schedules, or quality of the product, reports are sent to the production planners to keep them advised and to solicit assistance if it be necessary. These feedbacks are shown diagramatically in Figure 9.2.

As discussed in Chapter 5, the development of product designs may well require experimental work on models or prototype machines, which must in turn be produced in a machine shop or the equivalent. It is important to protect new product concepts from public disclosure for two reasons: to avoid disclosure to competitors and to preserve patentability of inventions. For these reasons, models and prototypes are usually made in-house. Fashioning these devices is a production function just as much as is the creation of the final product, although it may be done in a shop facility separate from that which makes the main product line. Nevertheless, it is shown in the functional diagrams such as Figures 3.2 and 9.2 as part of the output of the Production Function.

In a similar manner, the shop may produce special tooling such as jigs, fixtures, gauges, forms, templates, and the like for its own use, rather than procuring these from outside sources. The responsibility for providing these resources lies with the Resource Function, but that function does not exercise a materials conversion function. It therefore requests the procurement of the necessary materials for these resources and turns it over to the shop along with designs for the tooling, etc., furnished by the production planning people. This was discussed in Chapter 10. In some instances, complete machines may be built in-house. Construction of machinery and tooling is customarily conducted in a separate dedicated shop facility.

The diversity and complexity of the Production Function has been discussed in the second section of Chapter 6 in general terms. It is now time to examine the detailed structure of this function. Figure 12.1 is the expansion of Box 6 of Figure 6.5, labeled there as "Convert Material to Products."

## PRODUCTION CONTROL

The focal point in the conversion of materials into finished products is the Production Control Department. It makes the detailed assignment schedule of all the myriad of separate production tasks to the available resources; it tracks the flow of work from task to task, and determines priorities of work at each work site (machine tool or other piece of equipment) and it causes the delivery to each work site of all the needed items listed in Chapter 10. Its most important function is juggling the schedules so that all the many components of the final product are produced and delivered when they are needed with a minimum of overall delay and the highest possible production efficiency. It does this for all projects in process at any given time. This is no mean task!

This pivotal position is visually demonstrated in Figure 12.1, which

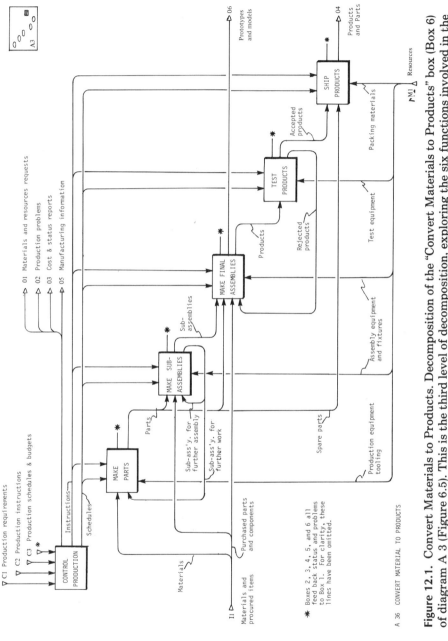

**Figure 12.1.** Convert Materials to Products. Decomposition of the "Convert Materials to Products" box (Box 6) of diagram A 3 (Figure 6.5). This is the third level of decomposition, exploring the six functions involved in the actual working of materials into products. It may also produce tooling for its own use, and prototypes and models for experimental work in the development functions.

136

is the expansion of the single Box 6 of its parent diagram, Figure 6.5; note that Figure 12.1 has the same inputs, controls, outputs, and mechanisms as are shown for Box 6 of the parent diagram.

The second control (C2) is comprised of all the production instructions generated by the manufacturing engineers in Production Planning. As described in Chapter 9, they will have expanded the detail design package into its component parts, and listed them all in the final Bill of Materials; they will have listed all the operations to be performed to produce each part and designed and ordered all the tooling and materials needed; they will have provided operating instructions and diagrams so the machine operator is instructed exactly what to do. In effect then, each machine operation is a self-contained task of work, with times and cost estimates included. In even a simple product there will be hundreds of these unit tasks to be performed.

Production Planning will have estimated the time required for each task. Because the tasks are, in general, sequential, the collective times will determine the total elapsed time needed to finish the fabrication of each part. However simply adding the task times, plus, of course, the times required for movement between machine sites, yields an overall time that presumes there is a machine or a truck ready to perform each task or move as soon as the work in process gets to it; this is called "infinite capacity" scheduling. Obviously, no shop has infinite capacity. It is the task of Production Control to reduce these early estimates to a schedule which takes the real capacity into account.

As pointed out in Chapter 6, there are four patterns of operation in industry, three of which are commonly found in the discrete parts industries:

Purchasing is a sequential pattern of order placement, waiting for delivery, acceptance procedures, and incoming stock inventory. Because many different items have to be procured, purchasing patterns proceed concurrently and in parallel.

Fabrication is also a parallel array of sequential operation patterns. Because the elapsed time for the manufacture of any given part varies widely, short flow time items can be included either early or late in the total calendar time span, at the convenience of the planner.

Finished parts, when turned into storage, follow a locational pattern.

Assembly follows a combinative pattern. Note that the total assembly operation may overlap the fabrication activities in time, and may in itself contain some short sequential fabrication operations.

Testing, run-in, and packaging all follow the sequential pattern.
Finished goods stores follow the locational pattern.
Shipping activities follow the sequential pattern.

Production Planning will have two fixed dates: today, and the scheduled completion date for a specific project. Beginning with the completion date and working backward in time, the planners can compile the time sequence for the total production and arrive at the necessary starting date. Figure 12.2 illustrates the compilation of such a date and the determination of the lead time required. Purchase of materials, fabrication, and inspection of parts follow a sequential pattern, in which there may be included some storage times. Part fabrications normally proceed concurrently. All the necessary parts must however be finished and ready for assembly when any given subassembly, or the final assembly, is scheduled to begin. Subassembly sequences are frequently carried out concurrently. The total elapsed time through final assembly, test, and packing is taken from the longest sequence of time spans.

It usually happens that the necessary lead time, reckoned backwards from the necessary delivery date, shows that the last possible

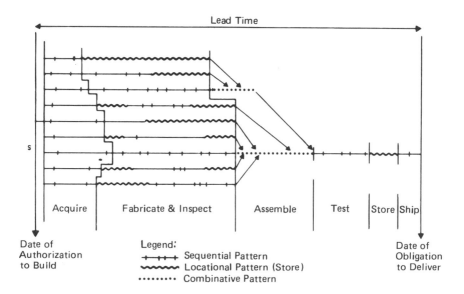

**Figure 12.2.** Production timing chart. Schematic sequence of manufacturing activities for one product. The scheduler determines the lead time required, between authorization to build and obligated delivery date, by adding the unit times in the sequence of the slowest component *S*.

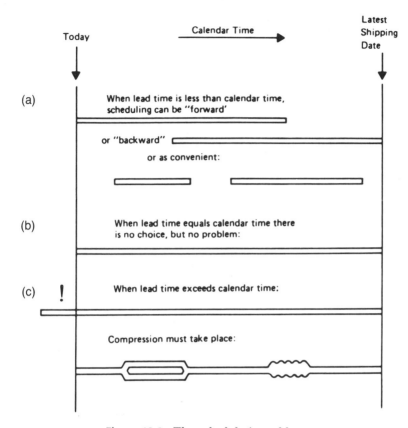

**Figure 12.3.** The scheduler's problems.

starting date has already passed! Figure 12.3 presents this situation diagrammatically. A series of recalculations and compromises to reduce the total necessary time must then take place:

Work may be farmed out to subcontractors.

Overtime production may be scheduled to increase the shop's productivity or a second shift may even be added.

Parts may be purchased from outside vendors rather than produced in the shop, thus releasing shop facilities for other parts that cannot be purchased outside.

Delivery of purchased materials may be accelerated by competitive bidding among sources of supply, or by placing a premium on early delivery from some reliable source.

These alternatives will have been examined and adopted when necessary by Production Planning.

Production Control may then be called upon to shorten the needed flow time for some key parts. For example, batches of parts may be split so that when a portion of the batch is finished on one machine it may proceed directly to the next machine without waiting for the rest of the batch. Alternatively, a batch of parts may be split up and routed concurrently over two sets of parallel and equivalent machines. Figure 12.4 diagrams the time savings accomplished by such control procedures. Another method of expediting a key part is to give it special attention, moving it immediately upon completion at one machine to the next operation, and giving it priority over all other parts waiting in queue for that machine. This of course introduces extra cost for the expediter, for the materials handler, and for the delay to the schedules of other parts in process.

The dilemma of trying to compress production time for some part is illustrated in Figure 12.4. However, this diagram relates to one part of one project; Production Control has many projects in the shop at any one time, and has the added problem of coping with each without detriment to any other. When all else fails, the completion date of one or

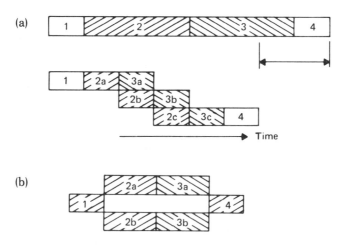

**Figure 12.4.** Reducing elapsed time in fabrication. (a) When sequences have very long operations, as 2 and 3, batches may be split into parts *a, b, c,* and overlapped in time, as illustrated in (a). (b) When several competent equipments are available simultaneously, batches may be split into parts *a* and *b* and processed concurrently.

more projects must slip, and warning of that problem must be sent back upstream to apprise all concerned. This warning feedback is shown in Figure 9.2.

Once the production schedule and the production plans have been agreed upon, Production Control issues the necessary orders to cause the actual production to take place. At this point in the narrative of manufacturing it becomes necessary to note that there are many different production control procedures, each tailored to the type of product being made, the size of the shop, and the volume of production. A relatively complex situation will be used as a prototype; those operating in less complex situations may count themselves lucky. Let us assume that we are part of an enterprise making a variety of products. Some of these are produced in a continuous stream of units, while others are produced in batches of units, at either regular or random intervals. All outstanding product units of the enterprise must be maintained with spare or repair parts upon demand. There is a continuing stream of new product development and research underway, requiring models and prototypes from time to time. Most component parts of the products are made in-house, but standard parts, hardware, and many components may be purchased from outside vendors. All of this is going on concurrently.

Every day a number of directives to produce some desired item will arrive at the Production Control Department. Let us trace one such directive and its ensuing activities. A group of parts is to be made and assembled into a unit, to be delivered for shipment. For the purpose of keeping track of costs, a work order number has been assigned to the directive. It must be presumed that the directive has taken cognizance of any design changes (ECOs) that are to be incorporated in the item. The directive will include a bill of materials that will list each part of the product, whether made in-house or purchased, and the number required. Production Planning will have already written production instructions and assembly instructions for every part; they will be on file and available. If this product is one which has been made before, the necessary tooling, fixtures, and gauges will have been made and should be in storage. (If it is a new design, Production Resources will have procured the necessary resources and have them ready. Let us assume that the example directive is for a product that has been made in the past.)

What does our production controller do? First he will look to see if there are any of the required parts in stock that are not obligated to some other work order. In an on-going business it is usual to produce certain frequently used parts in economical lot sizes, and therefore to have a stock as a source for use in assembly or for shipment as replacements or repairs. Standard hardware is usually bought in economical lot sizes and stored for any legitimate use. Therefore, a search through the inventory records will tell how many of each part are already available, and therefore the net production or procurement requirements may be reduced accordingly.

Knowing the total number of each part to be produced, the next step is to set economical lot sizes for the production, and to determine the time when each lot must be entered on the shop floor to assure assembly of the product at the due date.

If a part is used frequently in small quantities, it is common practice to keep a stock of parts available so that requests may be filled promptly. The economic production lot size is determined by a balance between two costs: Because the cost of setup is the same whether one or many parts are to be made in a batch, it follows that the larger the batch, the less the per part cost will be. This factor favors large batches. On the other hand, the interest on the investment represented by the finished stock, and the cost of storage, increase as the batch size increases. This factor favors small batches. The economic lot size will minimize the total cost. The standard handbooks explore this calculation.

Reference to the operations list for each part will tell what machine tools will be involved in its manufacture, and the set up time and cycle time for each. Multiplying by the number of parts to be made, the production controller learns how long each machine will be tied up with each operation in the sequence. He then checks with the work load already assigned to that machine in the time interval when he expects the batch of parts to arrive at the machine. On the day in question, is the machine tool already booked solid, or does it have time available? This is determined by reference to a work loading schedule for every tool and facility in the plant, a computer exercise usually run once a day. If time is available, it is booked. If not, he must search earlier or later dates for availability, or seek an alternate method of production.

Each operation on each part to be manufactured will then be scheduled for a specific day. At each machine tool, the tools, fixtures, gauges, etc., will be delivered from their respective storage locations. The list of all these items will be found in the operations list for the part, as discussed in Chapter 9. At the time of the first operation the raw material will be delivered from raw stock inventory, and at subsequent operations the batch of partially finished parts from the previous operation will be delivered. The moves will be initiated by the issuance of a work order by Production Control, sometimes called a "work release," which also tells to what account to charge the time and materials.

Work on an operation on a batch of parts cannot, obviously, be started until all the essentials are present at the machine, as pointed out in Chapter 10. It is the job of Production Control to see that this occurs, and to (preferably) receive from the shop floor an acknowledgment that they have the necessary elements to commence work. Production Control is later notified when the operation is complete, either automatically by a shop floor data collection system, or by the shop floor supervisor, so that the next series of moves may be initiated: all of the tools, fixtures, gauges, etc., must be sent back to their respective storage points (unless some of them are stored at the machine tool site) and the batch of parts must be started on its way to the next machine tool site. Then the whole process is repeated for the next operation listed, until the batch of parts is completed.

Quite obviously, there is a tremendous amount of data communication in this part of the enterprise, and the smooth operation of the shop requires that the data handling be accurate, timely, and as effortless as possible. In recent years, computer terminals and computer networks have been introduced to automate and expedite this task.

If finished parts are completed before they are actually needed for assembly, they may be stored in "Finished Parts Stores." Purchased parts are similarly stored in a warehouse as described in Chapter 11. In industries practicing continuous production, such as the automotive or the household appliance industries, finished parts may be fed directly to the assembly area at a controlled rate, thus eliminating the finished parts stores except for a small buffer store to ensure continuous operation of the assembly line even though some part production lines are temporarily stopped.

When the time comes to begin assembly, Production Control will issue a "kit list" or "pick list," directing the materials handling people to go to the respective storage areas and withdraw the correct number of each of the component parts for the first level subassembly, and to deliver them to the assembly work area.

At the appropriate point in the assembly sequence, the product will be moved to a test area where its performance is tested against standards set up by Development and accepted by the Project Manager and by Marketing. Ultimately, the accepted products are moved to the packing and shipping area, either for temporary storage in "Finished Goods Inventory" or dispatched to the ultimate buyer. All of these events are triggered by orders from Production Control, and are monitored by them.

In the preceding few pages we have been tracing the control steps necessary for the production of a batch of parts and the assembly of several such batches into a product. However, it must not be forgotten that there are other product orders, large and small, working their way through the facility at the same time and contending for the use of the same machine tools, operators, floor space, and resources of the shop. These all have finite limited capacities. As each new work order is entered, it must elbow its way into the work load, adding its burden to some parts of the system as it does so. As each operation and each assembly is completed, the facilities are unburdened by that amount. The task of Production Control is to reduce to a minimum the delays and inconveniences that result when some new work task overburdens some part of the facility.

Production Control may send to Materials Procurement, shown in Figure 9.2, Box 5, a request for more materials, as for example when Quality Control rejects a part and a replacement has to be made. Similarly, Production Control may send to Resource Procurement, Figure 9.2, Box 4, a request for replacement tools should one be broken or dulled, or for additional tooling or machinery.

Production Control must continuously apprise the budgeting and scheduling people (Figure 9.2, Box 2) of the status and cost of the work in progress, in order that they too may monitor progress, although their interest is not so detailed a surveillance as is the follow up of Production Control. Finally, problems which arise in the course of production that require a change in the production planning, or even in the design of the parts, must be referred back to Production Planning, Box 3. This feedback of production problems is also shown in Figure 9.2.

## MAKING THE INDIVIDUAL PARTS

We come at last, in this long narrative, to the actual conversion of input materials into finished parts—the one act of many that symbolizes the meaning of manufacturing in the popular concept of that term. Following the approach taken in the preceding pages, we will discuss the part-making task in the context of the execution of a work order to perform one operation of all those listed on the operation sheet, for one of all the parts listed on the bill of materials, for one end-product of all those in process at any given moment. This atom of the production world may be coupled to many other similar atoms to make up the total scene.

The operations list will assign the task to a specific work station—a machine tool and its operator, or to a workman if the task is a manual operation. If there are alternative equivalent work stations available, the local supervisor will designate which station and operator is to do the job. There may be only one part on which to work, as in the case of small projects or model work, or there may be a great many to be made, as in the minting of coins. Production Control will have seen to it that the input material and the necessary tooling be delivered to the work station, as described in the previous section of this chapter.

It would be ideal if the materials and the tooling arrived just as the machine tool and operator became available to work on them. If this cannot be arranged, it is usually desirable to have the material and tooling on site ahead of the time when the machine will be available to work on the task. This follows from the relative cost of having the machine tool idle, waiting to go to work as soon as the materials arrive vs. the cost of having the work-in-process sitting idle, waiting for the tool to become available. The work-in-process would have to be very valuable indeed to outweigh the capital cost of large modern machinery. However for simple inexpensive tools, such as arbor presses or bench grinders, the reverse may be true.

The first step is to set up the machine tool. This means that the machine, normally a general purpose device, must be equipped with holding fixtures and working tools specifically adapted to the work pieces to be operated upon. For example, a die casting machine must have the dies set in place and aligned; a printing press must have the plates locked onto the cylinder or platen; an automatic screw machine must have the cam bank installed and the cutting tools clamped in the holders; a machining center must have fixtures clamped to its work table. So equipped, the general purpose machine becomes a special purpose machine adapted to make one specific part or parts.

So equipped, a machine tool can make one or thousands of parts. Obviously, the set up cost will loom very large in the total cost picture if only one part is to be made, and will dwindle into insignificance if the tool is to run for months to produce many parts. In the automotive factories it has been economical to build and install special purpose machines known as transfer line machines, which completely shape a complex part such as a cylinder block. Once set up, such a machine may run for a 9-month period. However, whether the task calls for one part or for millions, some set up work must be done.

Next the operator must arrange to feed the input material to the machine. In some cases known as "first operations," the incoming material may be in bulk form, such as molten metal for the die caster, or rolls of paper for the newspaper press, or bars of steel for the automatic screw machine. The machine itself will advance the material for each successive part to be made. In other operations, semifinished parts will be the input to the machine, so that they are separate pieces. If they have been retrieved from the previous operation in an ordered array, such as a magazine or rack, and transported in that array, then the feeding of a single part to the machine tool is relatively simple, and may be automated. In such cases it is easy to segregate and orient one part when it is to be fed. If, however, the parts at the output end of the previous operation were simply tumbled together in a box or tote pan, feeding the machine tool is much more complicated. Individual parts have to be segregated from the mass, because one at a time is required. Sometimes this is easy to automate; other times it requires human manipulation of every part. Consider if you will a problem I once had: to devise a feeder for separating one fish hook from another after masses of them had been through plating and cleaning operations!

When a part reaches completion in a machine operation, it is in a known orientation. If it is simply dropped it loses its orientation; the entropy of the system increases, to use a term from thermodynamic theory. An effort is needed to reorient it for the next operation. This wasted effort can be avoided if production is designed as a system, and not as a disjointed sequence of isolated events. It should be avoided.

Frequently overlooked is the quality check the operator gives every part as it is picked up and fed to the machine, or to position the bulk material in the machine's input device. The operator will detect and reject any obviously malformed or defective part before it has a chance to clog up the mechanism. Sometimes an inspection operation of more precise nature is included in the operating instructions, or reliance may be placed on the quality control check that was done after the previous operation.

Orientation of the segregated part is the next task. There are eight ways to orient a 2 in.×2 in. photographic slide in a slide projector; only one of them is correct. There may be many ways to orient a work piece being fed to a machine tool, but only one of them is correct, and that one must be found. Automatic feeders working out of a mazagine receive the part correctly oriented by the magazine. Automatic feeders working from a mass of jumbled parts have to do the orienting job themselves. For manually fed machines, the operator does this task. The human operator may even be replaced by a robot with television eyes and artificial intelligence, if one can afford such equipment vis-á-vis the human. Once segregated and oriented, the part is placed on the machine and positioned. Figure 12.5 diagrams the time relationships of these acts.

Positioning proceeds in two stages. The part is first roughly posi-

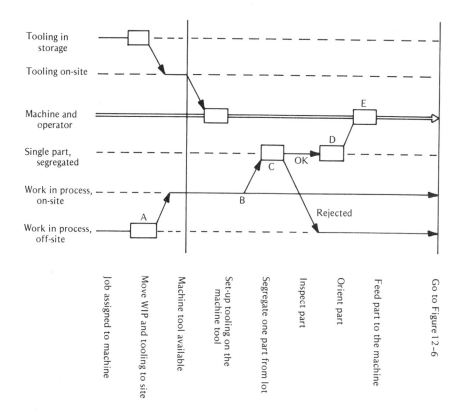

A  Omit if material moves by continuous feed (pipe, belt, conveyor line, etc.).
B  Omit if batch has only one piece.
C  Omit if prior inspection is acceptable, or the material condition as received is acceptable.
D  Omit if individual parts are presented in oriented array, or are bulk fed from rolls, reels, etc.
E  Presentation to machine's grasp may be by hand, the machine's own mechanisms, or by robot.

**Figure 12.5.** Set-up operations. Sequence of events, but not time spans, are indicated on the horizontal axis. Locations relative to the machine site are indicated on the vertical axis.

tioned, then precisely located. What this means in dimensions depends, of course, upon the size of the part. If the part is to be held by the closing of the fixture on the machine, then it must be within the open jaws of the chuck or fixture and near enough to the correct position so that the closing of the clamps will complete the location without damaging the part. If it is the size of a typewriter frame, let us say, then location within 1/32 in. would be a reasonable rough position; if it is the size of the balance wheel in a wrist watch, then a few thousandths of an inch would be a good rough position tolerance. Precise location depends upon the tolerance of the dimension to be produced in the machine. If a rough-turned shaft is to be mounted in a grinding machine for final grinding to a tolerance of a few ten-thousandths, then it will be positioned within a similar degree of concentricity relative to its reference surfaces.

Once precisely located, the part is clamped in place by the holding fixture or the chuck on the tool. However the part is held, the method must secure the part from movement, against the forces brought to bear upon it by the cutting or forming tools, or any other force from the operation. Appropriate tooling design is the responsibility of the tooling designers in the Resources function, as discussed in Chapter 10.

The operator is now ready to proceed with actual production. The operator may get approval from the supervisor before starting the first piece operation, or he may make one piece and get approval before starting on the remainder of the batch, if there are more to be made. The operation is repeated on each part in the batch – segregating, orienting, locating, precise positioning, clamping, cutting or forming, and unclamping. The finished part is removed to a rack, magazine, bin, or the equivalent. Sometimes the operator removes the finished part, sometimes the machine itself automatically ejects the finished part. Figure 12.6 diagrams the time relationships of the events in the actual operation.

In material removal type machining, there will be chips produced which must be removed from the work area; and other types of machines produce other types of offal that must also be removed from the scene. If the quantity of scrap is large, chip conveyors or the like move it automatically to a holding area for salvage. At the end of the work shift, the operator cleans his machine of any remaining scrap material.

To maintain quality control, finished work is inspected. This operation is a separate item in the operations list, but it may be interspersed with the machine tool operations. It may take place after the first part only, or after every part, or after every so many parts, or after the first and last part. If the machine is intended to be run unattended, as in an unmanned factory shift, the machine will probably be designed to make regular quality checks, and to react properly to off-standard production. If, for example, the diameter of a turned part is increasing, tool wear is indicated, and the machine may self-adjust its tool to correct before the tolerance limit is exceeded. Alternatively, the machine may shut itself down when it makes one, or two or three, bad parts.

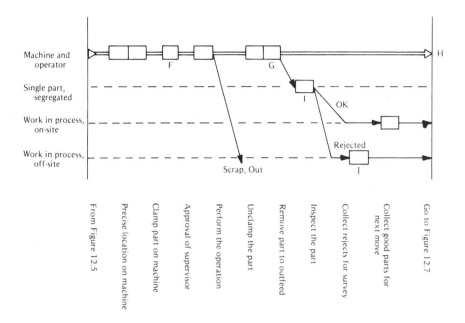

F  May be omitted, or performed after G.

G  May be automatic, manual, or by robot. In any case, orientation may be retained or lost.

H  If more parts remain in the batch, return to B of Figure 12-5 and continue. Otherwise, go to K of Figure 12-7.

I  May be done either for: a) none of the parts; b) the first part only; c) the first and every x parts; d) the first and last parts;  or e) every part.

J  After a preset number of rejects, may cause the machine to stop and signal for the operator's intervention.

**Figure 12.6.** Production operations. Sequence of events, but not time spans, are indicated on the horizontal axis. Locations relative to the machine site are indicated on the vertical axis.

Inspection may be required before the setup is broken down, so that if some parts are rejected, they may be reworked or replaced without another set up cost. Unacceptable parts are either reworked or scrapped. If they are scrapped, replacement parts may be required to keep the assembly date. When some scrap is normally expected, the batch may initially contain a few extra pieces so that the required number will be completed; if any of these extra pieces survive the whole process, they are saved for spare parts. In any event, the Production Controller is informed so that he may adjust his schedules accordingly.

Finally, the operator breaks down the setup machine tool. Fixtures and tooling are removed and released for return to their respective storage areas. The finished parts from this operation are released for transportation to the next operation on the list. The machine tool is cleaned and reported as available for its next assignment. All these moves and reassignments are triggered by the report to Production Control that the task has been completed, with the number of successful parts reported by Quality Control. Figure 12.7 diagrams the events of the break down sequence.

Any problem in connection with this activity will be reported back to Production Control (see Figure 12.1), and if necessary, they may in turn send the problem report back to Production Planning (see Figure 6.4). Such problem subjects may include machine breakdown, operator absence, tool failure or wear, defective material, excessive scrap, and the like.

All of the above is repeated for every operation on the operations list of every part in the bill of materials of every product. These are the "atoms" referred to earlier, which taken together, produce the parts needed for one product.

Finished parts may be stored temporarily in "Finished Goods Stores," to be dispensed later in kit form for assembly, or piece by piece for shipment to the field as spare or repair parts. Alternatively, they may be fed in batches or in a uniform continuous stream to the assembly function. When working in batch mode, the last operation on the last part in a batch ideally would be completed just before one of those parts is needed on the assembly floor. However, if the continuity of production depends on such close timing, and something does go wrong, the whole assembly operation and all its tributary operations come to a grinding halt. Something close to this ideal can be achieved, but deference must be paid to Murphy's Law, which says, "If something can go wrong, it will."

One of the things that can go wrong is discovering that there has been an error in the part production operation, and some or all of a batch of parts is unacceptable. If the part production setup has been broken down and all the tooling returned to storage, considerable delay and waste will be incurred in getting set up to correct the error. It is therefore wise to have the Quality Control operation take place as soon after part fabrication as possible.

Once again we see the importance of data communication in the control of operations in the Production function. Automated or computerized data processing abilities are essential to economic and expeditious manufacturing.

With some modifications, all of this section applies to any form of

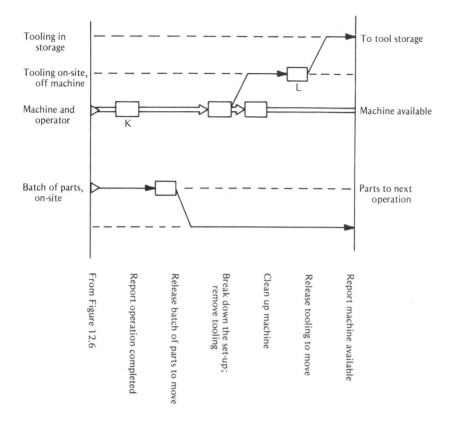

K May be reported manually or through automatic shop floor data collection system.
L Tooling may be stored on-site or returned to tool storage areas.

**Figure 12.7.** Breakdown operations. Sequence of events but not time spans are indicated on the horizontal axis. Locations relative to the machine site are indicated on the vertical axis.

manufacturing regardless of the product or the process. The reader must be relied upon to make the interpretation, if he is not in the metalworking field. It will not be too hard to do.

## SUBASSEMBLY OF PARTS

Assembly, except for very simple products, usually proceeds in stages, the first being a low level subassembly, and the last being the final assembly. Figure 12.1 is the expansion of Box 6 of Figure 6.5, and is entitled "Convert Material to Products." It shows two assembly functions in Boxes 3 and 4, the first labeled "Make Subassemblies" and the second labeled "Make Final Assemblies." The output of the first is shown feeding the input of the second, but a second output line has been drawn to show the recirculation of low level subassemblies into higher level subassemblies, an almost standard procedure.

The meaning of "level" of assembly may best be understood by reference to the bill of materials, described in Chapter 5 and illustrated in Figure 5.5. In the hierarchical form of the bill, the amount of indentation of an entry is indicative of its "level" in the hierarchy. The lowest level of assembly at any one point is indicated by the fact that all following items are indented one more space, but all are individual parts. When put together and fastened they take on the stature of an entity called a subassembly. It, in turn, joins its neighbors in the bill with the same indentation, be they other subassemblies, parts, or fastening agents, to be put together into the next higher level of subassembly, and so on until the last or highest level is reached, the final assembly. Assembly may be thought of as having a reverse tree structure.

The input to Box 3 is a collection of parts that have been made in the factory (Box 2), plus the purchased components acquired by the Material Procurement function. Production Planning will have written an assembly phasing or sequence instruction and forwarded it through Production Control to the assembly department. A start date and a due date will have been assigned for each level of the assembly work. A kit of parts, both those made in-house and those purchased from outside, will be drawn from the finished parts stores and delivered to the assembly spot. The delivery of the work order for each assembly task will also trigger the delivery to the assembly department of any fixtures and special tools necessary to perform the work; they will come from the store rooms of the Resource function.

Subassembly work is frequently done in a batch mode; that means that a group of identical subassemblies will be made at one time. Kit picking for batch mode assemblies is more efficient, because $X$ parts may be pulled from each part storage receptable at one time, instead of going back $X$ times to pull one part. If the subassemblies are not too large it may be convenient to lay them all out in a row on a bench, or on the floor, permitting the assembler to go from one to another, performing the same assembly step on each. Hence we have another term for the procedure, "bench assembly."

The operator performs a setup operation identical in principle to that described in the previous section on making parts. Figure 12.5

would apply if the term "machine" were changed to "assembly station," and many lots of parts were shown converging on the site. Figure 12.6 would also apply if the "operation" is understood to mean "assemble" and that it is repeated as many times as there are different parts in the parts list of the subassembly. Fixtures may be used to aid in positioning the several parts in their correct relative positions, and fastening devices are used to hold the parts together, rather than cutting or forming them.

As noted in Chapter 9, it may be necessary to perform some machining operations on a completed subassembly before it is ready to move to the next level of assembly. It is convenient in such cases that a new part identity be assigned, and that the operation sheet for that part number may direct it through fabrication operations before returning it to the assembly department. This option is also shown on Figure 12.1 by an arrow feeding back from the output of Box 3 to the input of Box 2.

The operator may be required to obtain approval from his supervisor or from the quality control inspector before proceeding to work. It may also be necessary to obtain an inspector's approval part way through the assembly task, in the event that subsequent assembly would forever preclude examination or, if necessary, correction of the assembly. For example, the encapsulation of an electronic component would prevent reworking if a later test showed that to be necessary. The final stage of the subassembly may well be inspected, either visually or by a functional exercise of the mechanism insofar as that is feasible.

Accepted subassemblies move on to the next level of assembly. Rejected subassemblies may have to be disassembled and the malfunctioning parts replaced or corrected; they are again inspected. If the subassembly cannot be repaired it is rejected permanently and referred to salvage. A dangerous alternative is to accept a slightly off-standard subassembly, making compensating adjustments in other parts of the final product. Such anomalies may cause problems later when some service man is trying to repair the product in the field and finds that he is dealing with a nonstandard thing. In any event, Production Control must be alerted to the state of affairs so that schedules may be corrected to show the true status.

Production Control is kept apprised of the status of the task. Recurrent problems in assembly may be referred to Production Planning for adjustment of the procedures, or even to Detailed Design to get a modification of the part designs to alleviate the problem.

When a subassembly task is completed, the operator breaks down the setup and returns fixtures and tooling to their proper storage area, and releases the finished work to the next station.

## FINAL ASSEMBLY

Final assembly is usually done one product unit at a time, as in automobile, aircraft, or machinery factories. Smaller products may be put through final assembly in small batches. In mass production enterprises, the final assembly may be done on moving conveyor lines, where pallets carry the growing assembly past successive stations where an operator adds a designated group of parts to each unit as it passes. The appropriate fixtures and tooling remain at that station permanently, and the supply of parts and subassemblies may be delivered there from a tributary subassembly line. In enterprises making products in smaller batches or at scattered points in time, or when the product is very large, as in steel rolling mills, the assembly work is completed at one station.

In any event, the operator makes the setup for his work just as described under subassembly in the previous section, and the final kit list of parts and subassemblies needed will be delivered in response to the work order from Production Control.

Final assembly operations are characterized by frequent inspection or tests, as the work proceeds, including in some cases the actual operation of the product to adjust it to its full operating capability. As final assembly approaches completion, these adjustment trials and tests approach the status of a final test for acceptance. If the product is a printing press, paper, ink, and type will be introduced and the press operated. If it is a computer, power will be attached, software installed, and test problems run. If it is a domestic hot water heater, water will be connected and the heater tested for leaks.

Some products are customized individually for specific buyers, as in the case of automobiles. The customizing modules will be assembled to the basic product in the course of the final stages of assembly, and similarly checked. Customizing in this fashion requires the allocation of a specific serial number to each individual product, and recordkeeping to assure that the customization is as ordered, and that that unit gets to the buyer who ordered it.

If at any point the product fails to perform as designed, the assembler may be required to take it apart and replace any defective parts or realign the components, and then reassemble the unit. Such work may cause a request to the finished parts storage for replacement parts over and above the original kit of parts. Inventory records will be adjusted, and the reason for failure of the parts in question investigated to determine whether or not changes should be requested in parts production or design. Reports of such nature go back to Production Control, who may refer them to Quality Assurance, to Production Planning, or even to Detailed Design.

## PRODUCT TESTING

Finished products are tested before release to Packing and Shipping. This test differs from that discussed on the previous page in that it is conducted by another person or agency, representing Quality Control. Final testing procedures are as varied as the kinds of products manufactured, so that a detailed description cannot be offered here. There are however, certain observations that apply to all manufacturing fields.

Quality cannot be inspected into a product. Bad products can be culled out by inspection, but quality must be designed in at the product design stage, and planned in at the production planning stage. This is called "Quality Assurance," in distinction to "Quality Control," as discussed in Chapter 5. However it is achieved, the manufacturer is responsible under the law for the safety of the product in use, and under the warranty provisions of the sale for the adequate performance of the product in use. Testing is therefore of major importance in the manufacturing field. Human nature being what it is, it is usually deemed wise to have the final test administered by others than those who made the product; the latter always seem able to make their products perform, even though no one else can.

Frequently, in the case of large and expensive products, or one-of-a-kind products made to the buyer's order, it is common to have the buyer present for the final test and demonstration. If he does not then raise any objection to the product's performance, he is less well positioned to complain of it when he begins to use it. Testing of large products may very well take place on the final assembly floor, simply to save the effort of moving them an extra time.

Accepted products move from test to the finishing and shipping function. Products rejected by Testing must be returned to the jurisdiction of the Final Assembly function for disassembly and rework as necessary. This is shown in Figure 12.1 as the fifth output from Box 5. When it is necessary to send spare or repair parts to the field for support of products in use, the parts will be drawn from Finished Parts Stores, if there is a supply available, or a batch will be made, and sent to Shipping. Quality Control of these parts is maintained through the normal inspection at the point of fabrication.

As also shown in Figure 12.1, Testing reports status and problems back to Production Control.

## SHIPPING

Shipping is the last of the functions in the conversion of materials into products. It is comprised of a number of steps – finishing of the product, possible disassembly to achieve portability, protective packaging, temporary storage, and finally discharge to the transport medium.

Finishing the product includes a touch up of the surface paint, if necessary, or even painting a machine in the buyer's choice of colors; addition of the name plate with model and serial number, if there is such a thing; and the addition of any spare parts, manuals for operation, parts lists, and other documentation which may be called for. This operation is as varied as are the products manufactured.

Small products may be packed as units, or even as groups of units, but large machinery and the like may have to be taken apart even to get it out of the factory, on to a railroad car or truck, and into the buyer's plant. Obviously the design of the product will anticipate this need for disassembly before shipment, and joints in the electrical, plumbing, and shafting or other mechanisms for this purpose will be provided for at the break points. This disassembly will be done on the final assembly floor, under the supervision of that department.

Small products may be shipped in boxes of many sorts, properly cushioned if they are delicate, and sealed against tampering or pilferage. Rugged parts, such as a bundle of automobile tires or lengths of pipe, may simply be strapped to a pallet and shipped as is. Household washing machines will be set on a pallet and shrouded with a corrugated cardboard box to protect the surface. Machinery will be put on skids or platforms so that they may be picked by fork lift trucks or overhead cranes. Protection against weather during shipment will vary according to the product, the length of the trip, and the time of year. Perhaps the most protection will be needed on trans-oceanic deck shipments during the winter. Another function of packaging is to prevent theft or mischievous tinkering. A well covered product does not invite curiosity.

Some products will be loaded on the transport and dispatched as soon as they are packed. Others will be stored temporarily awaiting prepayment by, or credit clearance on, the buyer. Others will be held awaiting other products to achieve a bulk shipment and the economies thereof. Still other types of products are made in a steady stream and are stockpiled in "finished goods stores" until orders for them are received. A large inventory of finished goods is undesirable, because the investment in the products is at its maximum, and cannot be recovered by billing until there is a buyer.

## SUMMARY

Conversion of materials into finished products is the sixth and last of the functions which comprise Production. It receives product designs and production plans, production requirements and schedules and budgets, which govern its action. It receives materials to be converted into parts, and resources with which to do the conversion. It plans its work, fabricates individual parts, assembles them into products, tests the products, and ships them to users. It makes models and prototypes for the Development Function, and sometimes makes its own tooling and machinery. This is a complex function, varying widely in its details in accordance with the product line being made.

Conventionally, the shop, as this function is known colloquially, will be operating concurrently on tasks for a number of different products and product lines, in various stages of completion, and working under a variety of priorities. Production Control is the unit that plans the total load of the department, breaking the load up into work orders, distributing the work orders to the various machines and facilities, leveling the work load to maintain efficient and timely results, and monitoring the whole operation. When a work order is released to the shop, simultaneous move orders are issued to move the material to be worked upon, the fixtures, and the tooling to the machine site.

At the machine or assembly site, the operator sets up the machine tool with the fixtures and tooling, arranges the feed of the materials to the machine, and their removal when the machine operation is complete. Incoming material is checked for quality, and output parts from the machine may also be checked or inspected. The machine operation is repeated on every unit in a batch of parts. Scrap is removed and finished parts are moved to the next operation, or if fabrication is complete, to final inspection and finished parts stores.

Guided by the Bill of Materials, kits of parts from the finished parts stores and the purchased components inventory are pulled and sent to subassembly sites. Completed subassemblies may be sent back for more shop work, or forward to higher levels of assembly. Inspection is frequent during the total assembly process, and rejected assemblies are reworked and repaired. Final assembly completes the product, and final inspection and test completes the process. Accepted products are finished, packed for shipment, and stored until shipping is directed.

All components of the shop report status to Production Control, and may request assistance in solving problems requiring additional material or resources, or changes in detail designs of the parts, production procedures, or work scheduling.

# Chapter 13

# Product Support Function

The Product Support Function is the fourth major function within the Manufacturing Function of the corporate enterprise. A glance at Figure 3.2 will refresh the reader's memory on the relationship of these four primary-level divisions of the total manufacturing activity. It has two functions: first, it provides the Marketing Function with spare and repair parts, special equipment, and resources needed for product maintenance and operating and instruction manuals; and second, it feeds back information to the Design and the Production Function from the field from the point of view of the actual user of the product. This marketplace feedback is supplemental to the similar feedback from Marketing.

Support of a company's products in the field is an essential component of long term success for a manufacturing firm. Repeat sales to an old customer depends upon his satisfaction with the performance of a product, and with the support he receives during production difficulties. It also depends upon the expectation that the company will continue to do the same in the future. A good reputation will spread to other users in the field, and thus will contribute to enlargement of the market penetration. Also a user who is pleased with his purchases and the services he gets will be easier to deal with, when and if something does go wrong.

It is important and timely to differentiate clearly between the functions of Manufacturing and Marketing. Part of the marketing function is the perception of the need for products in the field served by the company, and the opportunities provided thereby to the company. In most fields of endeavor, the user's technology is in a constant state of evolution. Researchers develop new materials and new methods. Sources of supply of materials change with world wide economic and political situations. Worker skills and availability change with the political and social changes. Buyer's interests and demands are in a constant state of evolution under the influence of education and advertising. As a result, no field ever remains static for very long. It may be growing or shrinking, or just changing its nature, but it rarely remains stationary. It is a primary task of Marketing to perceive these changes and recommend a course of action to the company to take advantage of them. How should the company's product design specifications be changed? How should marketing policy be changed? Advice of this sort may be for merely a minor change in some design parameter, or for a major new product line introduction, or for anything in between.

Part of Marketing's function is to be alert to the marketplace, forward the sort of information described above to the corporate management, and offer advice on product policy. Part of Manufacturing's function is to receive this advice and incorporate appropriate changes in the designs and the products of the enterprise.

While Marketing is wholly responsible for the financial aspects of product sales – price, terms, credit, timing, etc., it has no responsibility or control over the details of the design and production of the product, except as an advisor. It also contributes information and insight regarding the technology and usages in the product user's field. It relies upon Manufacturing for products, parts, instructions, and maintenance equipment which it sells and distributes.

On the other hand, contact with the buyers and users of the company's products is the province of the Marketing function. Knowing in great detail the individual customers and their idiosyncrasies, and having dealt with them over the years, Marketing quite properly reserves to itself the prerogative of customer contacts and customer relations, and Manufacturing should accept Marketing's intermediation when field contacts are needed.

An underestimation of the marketing task is not implied by the briefness of these few preceding paragraphs. They underrepresent by far the magnitude of the total marketing function. Another volume as large as this could be devoted solely to marketing, but its contents are beyond the scope of this book. However, it is essential at least to touch on the

Marketing function, because it touches upon our field of principal interest, the Manufacturing function.

The function that is the subject of this chapter – Provision of Product Support – is one agency of interchange and cooperation between the Manufacturing function and the Marketing function. Its operations are controlled by a directive from the Project Administration, Figure 3.2. Its inputs are threefold: it gets information concerning the design of the product from Product Development, Figure 3.2, Box 2; as to how it is manufactured, from Production, Box 3; and it may receive purchased items or machine components for distribution.

It is essential that the Product Support function have access to the design information so that they may prepare operating and maintenance manuals for field use. Parts of the product known to be subject to wear or damage will be identified so that spare parts may be readied for quick response. Assembly configuration and sequence will be converted to "exploded" views for ready comprehension, and included in the manuals. Figure 13.1 is such an exploded view.

Production will supply information concerning the assembly and disassembly sequences, the nature of assembly fastenings, and similar information. Included will be information on the tooling and fixtures necessary for assembly and disassembly. For example, an automobile company will design jacks for lifting the cars to change tires, and designate where the jack is to be attached to the car to make the lift. Special wrenches for reaching otherwise inaccessible cap screws will be designed. If such items are needed, Product Support is responsible for perceiving the need, requesting the development, and arranging for production of the needed items.

Outputs from the Support Products function are of two sorts. The function prepares informational material concerning the products. This includes operating manuals, service instructions, maintenance and repair manuals, and a parts list or bill of materials that exactly defines the configuration of the specific machine or product, identified by serial number. This latter item is essential for large products made in small numbers, such as airplanes or steam turbines, whose detailed designs may change from one serial number to the next. Even automobiles may leave the factory with variations from one car to the next coming off the assembly line. Owners or service men will be reading the manuals and instructions as pertinent to one specific serially numbered unit of the product, and will want to know precisely what they have before them if they have to order repair parts or ask for help from the factory.

The bill of materials will be rewritten in a form containing only the facts pertinent to field use. Hardware such as nuts and bolts or

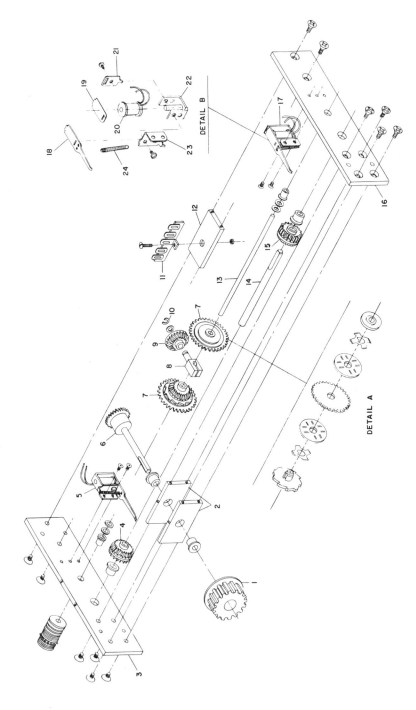

**Figure 13.1.** Sample of an exploded view drawing of a capstan drive.

DETAIL A

DETAIL B

162

electrical connectors will be described well enough so that a service man at a remote location can obtain local replacement for these commonly available parts without sending an order clear back to the factory.

The second sort of output from this function is the variety of tools, fixtures, and materials needed for field support of a product. The car jacks and special wrenches referred to above may be made by the factory, or made to order by a vendor or a subcontractor. In either case they will be supplied to the Marketing Function's Service Division for distribution, upon orders from the latter unit. "Materials" means anything that is used by the product in its operation and is not readily available in approved quality and quantities from other sources. Hydraulic oils, special lubricants, typewriter ribbons, and air filter cartridges are examples. These, too, will be procured in bulk quantities from suitable sources and made available to Marketing for distribution to the field.

Upon occasion a large and complicated product such as a machine tool or a loom may need more than just maintenance and service; it may have to be rebuilt and possibly modernized. Such activities may be performed by companies specializing in such rebuilding or reworking, or it may devolve upon the Service Division of Marketing. The product is taken out of action, disassembled, evaluated, and a plan is made for the updating. Replacement of major parts of the product may be necessary—parts that ordinarily would not be available in spare parts stores. Replacements will have to be made that will be the exact equivalents of the displaced worn parts. Here again, the importance of correct entries in the bill of materials for all the component parts of that specific serially numbered product will be apparent. Production Support will channel the orders for such parts and assistance to the proper spots in Production. Modernization as well as rebuilding may involve the assistance of the Development Function.

Finally, Product Support will feed back to the Development Function all knowledge gained relative to the performance of products in the field, and any hints as to the change in the technology of the users of the products, for their guidance in future development work. It will report unduly frequent failure of a part of the product, frequently requested minor changes in spare part designs, or apparently moribund items detected in parts usage records. Product Support will feed back to the Corporate Legal Function any knowledge concerning the safety and performance of the company's products which could affect the liability of the company.

## SUMMARY

The Product Support Function provides to the Marketing Function the spare and repair parts, the special equipment and resources needed for maintenance of the products in use in the field, and the operating, service and maintenance manuals. It also feeds back to other Manufacturing functions all information pertinent to better design, better production, and lessened liability for the company's products.

Product Support is a working point of contact between the two major parts of the whole enterprise – Manufacturing and Marketing – so it is appropriate here to differentiate clearly between the responsibilities and authorities of these two.

Marketing is solely and wholly responsible for establishing marketing policies, for advertising the products, selling them, installing and servicing them. Manufacturing is solely and wholly responsible for developing and producing the product line, and for provision of support for the company's products in the field. Its contacts with buyers and users of the company's products are conducted through the agency of Marketing. Product Support is one avenue of such contacts.

# Chapter 14

# Data Processing in Manufacturing

As manufacturing changes from an art to a science, the importance of information, of information manipulation, and of information management becomes more and more apparent. If we are to analyze this science of manufacturing to determine the rules which govern it, to use those rules to predict behavior, and then to learn what parameters we must measure in order to control the behavior, we must learn to cope with data processing.

This is not a new concept. Pythagoras, who lived and worked a little before 500 B.C., founded a school of philosophy and mathematics. Among other things he discovered the famous Pythagorean theorem regarding the relationship between the lengths of the sides of a right triangle. His school of philosophy taught that "all things are numbers," meaning that the essence of things was number, and that all relationships could be expressed numerically. This concept has had theoretical acceptance down through the ages, but insofar as practice is concerned, the manipulation of numbers was limited to very primitive mechanisms up until the last 200 years. This difficulty has somewhat obscured the basic premises of the Pythagoreans.

The technology of data processing has made prodigious strides in the last few decades (1950 to 1980 and beyond), and equipped us with powerful tools for handling information. In fact, the tools have so far changed the work of handling data as to change our whole perception of that task. It is important to see the elements of manufacturing through this new perspective.

We have dissected and examined the science of manufacturing in terms of functions, represented by boxes in our diagrams, and data or material flows between the functions, represented by arrows in our diagrams. The functions may generate or transform data; the flows only transfer the data. The use of this convention presupposes that any movement of either materials or information can be represented by a data movement. This is true.

Information may be embodied in words or numerals, representing concepts in our minds. Words are composed of letters, and letters may be represented not only by their graphic symbols but also by simpler bits, such as the dots and dashes of the Morse Code of telegraphy, or by the binary codes such as ASCII (American Standard Code for the Interchange of Information). The same applies to numbers (the mental concepts of quantity) and their symbolic forms – numerals. These too may be represented by the binary bits 0 and 1 of the ASCII or similar codes. These codes were expressly developed so that digital data processing equipment could store, retrieve, and manipulate the information. The records frequently take the form of magnetized spots on a tape or disk.

Another large segment of information is represented by geometric forms – the curve of an airplane wing or the hood of an automobile, the shape of a bowling pin, or the curved sweep of the cables of a suspension bridge. Interpreting geometric forms into numerical forms was solved by the classic Greek scholars two millennia ago, and is now common technology. Mathematical curves like the catenary of the suspension bridge cable can be expressed directly in numerical form. Non-mathematical forms, such as the bowling pin's shape, can be expressed by measuring in Cartesian coordinates the position of a succession of points on the curve, and announcing that the curved surface passes smoothly and successively through those points.

While it is frequently convenient to represent the shape of a part to be duplicated in a factory by a template which will control a copying lathe or a die sinking machine, this geometrical information may be reduced for more convenient storage and manipulation into the equivalent numerical form. Another familiar example of a physical embodiment of information is the bank of cams which control the intricate motions of an automatic screw machine. A dozen disk-shaped cams, mounted on a shaft, will control the motions of a dozen cam followers which in turn cause the lathe's tools to operate in sequence to produce a given shaped part. The cam bank may be removed and stored for future use. Later day versions of machine control by stored data are the numerically controlled machining centers and lathes, where digitized numerical data is prepared in advance of manufacture and stored, and then may be read by the machine's controller to cause the machine to perform as desired. The same data record may be read out into the machine as many times as desired, or stored for future use.

Pursuit of this line of reasoning will show that every facet of manufacturing may be expressed by information in some form. All the functions and information flows of our diagrams can be reinterpreted as either data generation, data transformation, or data transmission. For example, let us trace the progress of the design of a part, from conception to finished hardware: The designer has in his mind the concept of a part of a mechanism; that is a data set. He transforms it into a layout drawing. The layout is transmitted to a detailer, who transforms it into a set of detailed drawings. The detailed drawing is transmitted to a manufacturing engineer, who transforms it into an operation sheet, a material order, tool specifications, and instruction sheets. These are transmitted to various people, who transform the material, drawing, and instructions into a finished part. That finished part embodies in its shape and size the same set of data that was in the mind of the designer in the first place. The data set remains intact, in spite of the many transformations and transmissions necessary to convert the concept into a finished part.

To take another example from another area of manufacturing, let us consider the activity of a shop supervisor. As a manager the shop supervisor performs planning and control functions. He receives from his supervisor the assignment of work to be accomplished by his unit in the coming time period. This is in fact a data transmission. He expands the assignment, breaking down the total task into a series of subtasks, allocates them to workers at machines, and optimizes the overall operation of his unit by adroit assignment of tasks to the best fitted worker, in the sequence that will meet the total assignment schedule in the time available. This is in fact a data transformation. The shop supervisor then distributes his instructions to the workers. That is a data transmission.

These two examples are drawn from what is frequently considered to be two planes of manufacturing activity – one, the operations leading to the physical creation of an end product, and two, the management, timing and control of those operations. In reality, these are not separate planes of activity, in so far as the data is concerned. The data involved in manufacturing pass back and forth from the operation plane to the managerial plane with complete freedom. It is neither labeled as operation data or management data; there is no such distinction. For example, when a batch of parts has been completed, that operational fact generates a bit of data that goes back to the production controller for his next managerial orders. If it did not move thus, the managers would be operating in the dark. When the production controller decides on the priority to attach to some shop order, that managerial bit of data is transmitted to and controls the shop floor. If this type of data movement could not take place, the shop would be running out of control.

Reference was made above to these two environments as planes of activity. If they must be visualized as separate planes of activity, they should be regarded as the two sides of one plane, and a very permeable plane at that.

This discussion of data in manufacturing has treated it as abstract information, and so it is. For every communication of information, three things are essential: there must be a sender, a receiver, and a message; but to be useful the message must be visible and accessible to those who are to use it. When the user is a machine, as in the numerically controlled machining centers, the accessibility is achieved through a direct coaxial cable connected to the host computer's memory, or through the medium of a perforated paper tape. When the user is a person, the message must either be communicated orally, printed or written on a piece of paper, or displayed on some electronic device.

Oral communication is the oldest method of transmitting data. Like other primitive forms of communication, it suffers from the fact that what gets transmitted is what the speaker thinks is a fact, hopes is a fact, or wants his listener to think is a fact. When data are filtered through the human brain, they may or may not be improved. Sometimes the raw data which the cycle counter on a machine tool might display needs interpretation or evaluation to be truly useful. At other times prejudice, venality, or just plain error may creep into the interpretation.

The listener usually makes a record of the communication if the data are to be used later by him or by others; again the potential for human error is introduced. The ambient noise level in some factories can cause conversation to be difficult and prone to error. Language and dialect barriers exist between some persons, and of course there is the ever present hazard of simple misunderstanding. At times, people use idiomatic or ambiguous language that is subject to several interpretations. Voice communication of data is not recommended where accuracy is important. On the other hand, a loud shout of "FIRE" is superior to any amount of written memoranda on that subject.

Written communication of data is less subject to error, although the possibility of human error still exists. The same problems regarding the sender's perception and interpretation of what he wishes to communicate exist here as in oral communication. Written numbers and letters are less likely to be misunderstood than spoken words, and the recipient is less likely to introduce new errors in his turn. However there is an inherent drawback to written communication: it is a batch operation. The information is written, a copy is filed, and the original and other copies are forwarded by the mailing system, special messenger, or a conveyor system to the destination. There are time delays between the occurrence of the event that generated the data, the time of reporting, the time of delivery, and the time the data comes to the attention of the intended recipient. During all this time, written data have remained unchanged, even though the facts reported may have altered radically.

Printed forms have been devised by the thousands to expedite and to ease the burden of written communication. They vary widely, from simple lists which are merely checked off by an individual worker as it comes to him, to elaborate forms that elicit an entry for every conceivable aspect of events that might be important and reportable, even though some may be redundant or unnecessary. Every company has its own particular format for its paper work, and these forms get deeply embedded in the company's procedure. Changing them is difficult, even though the dead hand of the past makes some of the forms a liability rather than an asset. Experience shows that if they must be used, written communication forms should evolve and change with time and the changing circumstances of the company.

## COMMUNICATION VIA COMPUTER

The advent of data processing equipment, specifically the digital electronic mechanisms collectively and colloquially referred to as "computers," has changed much of manufacturing information communications. There are certain basic principles that must be remembered.

In manufacturing there are two classes of data with quite different characteristics. The distinction between these two classes of data is made because the internal format of the data—particularly geometric data—in a computer and its transfer and manipulation calls for different data processing capabilities. The first class is data which defines the product. This includes both geometric definitions and alpha-numeric definitions. A detailed drawing (see Figure 5.2) will contain both the pictorial or geometric elements, as well as the dimensions, material specifications, identifying numbers, etc., in alpha-numeric format. The bills of material (see Figure 5.5) contain solely alpha-numeric elements. With other similar documents they completely define the product – part by part and in the aggregate. The second class of data pertains to the control and management of the operations of the enterprise, regarding time, location, personnel, cost, and the like. This class is completely alpha-numeric.

Referring back to our diagrams, the functions depicted in the boxes will generate, transform, store, and use data. Some of it will be generated or transformed by the people manning the function, some of it will be processed automatically. For example, a designer may create a part design on the face of a cathode ray tube in his Computer Aided Design terminal (see Figure 5.4); in doing so the designer transforms a data set generated in his head to the computer processor. He may then find it necessary to make calculations which determine the necessary dimensions of the part to withstand the loads to be placed upon it. This is done in a large computer programmed with the stress or similar design algorithms. In so doing the designer will draw from the CAD terminal the dimensions of the part, and add the loads calculated elsewhere. The computations known as Finite Element Analysis are excellent examples of some computer aided engineering calculations which are so complicated that they could not be performed manually, but may now be accomplished by the aid of the large computers.

When a manufacturing engineer designs methods, writes operation sheets, and designs tooling, he too is generating data. He may start his work by calling up from the CAD terminal the original design data and extracting from it what he needs. If it can be transferred electronically, it is then unnecessary to recreate and reenter the design configuration in the manufacturing engineer's CAD terminal. He may call upon his iterative or his generative part programming system to help write the production instructions, or he may have to generate them himself. While he has the configuration of the part on his terminal, he can generate a control tape for the numerically controlled machining centers or lathes, again without reentering the data in his NC programming terminal.

When the part gets down to the stage of production, the Production Control engineer can transfer from the production instructions the necessary orders for materials, tooling, etc. Again, the tooling designer may transfer the part configuration to the design terminal and design around it the shape of the necessary holding fixture; the part outline is then erased and what is left is the drawing of the socket in the fixture.

Information may be generated automatically in a function. When an operation is performed on a machine, the fact of the completion is a newly generated bit of data. Many machine tools, such as punch presses or automatic lathes, have cycle counters on them to record the number of times the tool has operated. The movement of a batch of work-in-process from one machine site to another, or the completion of an acceptance test or inspection, are other types of events that of themselves generates a bit of data.

Local computers may be applied to solve the legitimate needs of the various functions we have discussed. Examples are legion: the computer can help with the generation and maintenance of various bills of materials, the generation of the Master Schedule, the creation of production instructions, the creation of NC tapes, the creation of tooling and material needs, the control of inventory, the purchasing function, the operation of the automatic storage and retrieval systems in the warehouse, testing, and in many more functions. The data generated in these operations are essential to the efficient and expeditious operation of the functions in which they are installed. However much of it is for purely internal use. A small amount – quantitatively – is of great value to others, particularly to the management functions, and may be made available for transmission to others who need it.

Finally, it is possible to arrange a local computer to monitor the work in its domain by receiving reports of accomplishment and comparing them to the plan or target established in advance by Production Control. When the performance deviates from that set up as the objective, the computer can alert the operator, or the supervisor, or can automatically halt the work of a machine or readjust its tool setting if the output is moving out of the acceptable tolerance zone. This is a local control, but its actions may be of interest to the production control unit, and may be reported to that unit.

Referring again to our diagrams, the arrows represent movement of data or material from function to function. If we accept the fact that material may be represented by data, this is an example of data transmission.

As mentioned above, some equipment has built-in sensors which note the progress of operation. These devices may be cycle counters on repetitive operation machines, or outfeed from the controllers on NC machines. These may be arranged to report by direct wire to Production Control on each such cycle. (Note, however, that not every cycle will produce a good part; inspection will have to determine how many good parts were made.) Other machines with longer cycle time, or tools working on single part batches, will be reported through shop floor data collection systems to the Production Controller. The operator, or his supervisor, may actuate the reporting mechanism. The same applies to hand operations where there is no machine involved. Many of the documents involved in manufacturing carry machine readable information which may be read by the electronic data processing equipment and put to immediate use without the risk of introducing human errors. Workers' identification cards, work orders, drawings, etc., may carry bar codes or magnetically encoded strips for these purposes. The operator inserts his ID card, the work ticket, etc. into the reporting device, keys in any additional data, and punches the "Transmit" button; the data are read from the document and transmitted directly to the production controller, who needs to know that information.

As more and more of the departments in a factory become equipped with minicomputers to handle their own problems, the opportunity to tie all these data generating points back into a single network is presented. Whereas in the past the source of data was human, and the use of data was by humans, the trend with computerization is toward the direct generation and utilization of data by local computer units. Data (messages) may then be moved from the source or sender directly to the user, and practically instantaneously. It is up-to-date at all times; none of the obsolescence referred to above with written communications will occur. Data may be made available wherever it is needed. The user may be a human, but is more likely to be some direct application to control of a function.

It is therefore unnecessary to have duplicate copies of information; redundancy is eliminated; the user goes back to the source when information is needed and as often as needed. This eliminates the source of errors in handwritten notes stored in a greasy notebook in the operator's or supervisor's hip pocket.

It follows that there should be only one source for any one bit of information – the department or agency that generated that information; and only the originator may have the right to change the data. This ensures that anyone needing information will receive the latest version; and because it is always and instantly available, there is no need to save a bit of data for later use. Stored data other than the original source is liable to be obsolete because of changes at the source, and should therefore be forbidden.

When computers are integrated into a system, the connections are either wires such as coaxial cables, or optical fiber links. Data move over these connectors in coded form, and with checks for accuracy of transmission, in accordance with standard protocols which need not concern the users. The mechanisms by which space and time are virtually eliminated in the process should be transparent to the user.

At one time it was routine to have the main frame computer print out every night a complete status report of every work order in the plant, the backlog of orders at every machine site, and similar "dumps" of gross, undigested information. Huge stacks of computer print out were delivered every morning to one and all. Little of it was ever used. If some of the bits of data in the stack were wanted, it would take a genius to find them and correlate them. Fortunately, this habit has abated. When a person needs some specific bit of information, he can ask the system to get it and he will get that bit, and nothing else. If he wants to see a summary or an analysis of performance, he can specify just what factors he wants considered, and how they are to be correlated. He can have this relatively brief read-out recorded and saved for reference in future similar analyses.

The variety of studies possible when data is available to a computer network is limitless. For example, a manager may ask to have displayed on a CRT for study:

All the products that use a certain part;
The backlog of work at any given work center;
The priority listing for all orders in a queue;
The total material and labor cost invested in a work order to date;
The record of good and bad parts from any given operation;
The amount of inventory of some kind of raw stock;
Absences and late arrival records for an individual;
Delivery performance of a vendor;
Location of any given work order or group of orders;

and many, many more such studies.

Because special purpose computers have been acquired by many different people, and to suit many and varied needs, it usually follows that a plant will have a wide variety of computer makes and models. Each will have its own control language, data structure, and operating system, and experience shows that rarely are they compatible; nor is it reasonable to expect that an enterprise will have a set of distributed computers that are completely homogeneous. The makers of computers keep inproving their product to meet the expanding demands of manufacturers, competing vigorously for the market. A heterogeneous computer environment is therefore to be expected.

The exchange of data between functions is essential, as it always has been, whether the exchange is manual or electronic. However if it is electronic, then the source computer, the message format, and the receiving computer must be compatible. To accomplish this it is essential to have what is called a neutral database. This means a database that is accepted equally well by all the computers in the system. It must be invariant over time and indifferent to the computer hardware and software. Further, it is desirable that this database and the communications protocol should be transparent to the user. When a manager asks the system when a product for some specific buyer will be ready for shipment, a date is needed; the manager does not want to have to tell the system where to find the bill of materials, where to find part production status, how to aggregate completion dates, and forecast assembly completion.

For managerial purposes it is possible to have the raw data representing manufacturing accomplishments collected over some specific period or periods such as an hour or a day, and analyzed in accordance with some prescribed procedure. Only the key elements of the analysis, or the summary, need be reported; the supervisor is not interested in the mathematics of achieving the results. Such analyses may be called out on a special occasion, or may be made routinely; they may even be arranged to alert the supervisors if some key criteria approach, or exceed, their prescribed limits.

Transformation of data may include such studies when they are used for simulation of possible future courses of action. When a manager has a number of possible courses of action open to him and must make a decision, he may ask the computer system to run an analysis several times, each with a different set of proposed conditions. If the computer logic is not capable of making a choice of the preferred conditions, he then can compare and select the optimum course of action. Such capabilities are called "Decision Support Systems."

Computerized information systems have thus contributed to the orderly and effective use of information in the management of manufacturing. Suitable safeguards may be built in so that such insightful analyses will be accessible only to those who have a need and a right to know about them.

As time goes on and information processing skills and facilities improve, manufacturing establishments will utilize their information handling capabilities to understand and to control the science of manufacturing. With understanding and control will come greater productivity and surer attainment of the corporate objectives.


## SUMMARY

A fundamental precept of this book has been the concept that all the facets of manufacturing can be expressed in terms of data generation, transformation, and transmission. All of the alphabetic and numeric forms of data can easily be handled in digital electronic data processing equipment. So too can the geometric forms encountered in manufacturing.

Data must be manipulated both in the operations involved in manufacturing, and in the management of those operations. Data move freely back and forth between these two planes of activity.

Information may be transmitted orally, in written form, and in electronic signals within a computer. The latter presents many advantages to the management of a manufacturing enterprise. There are certain virtures to computer data processing: Data are stored in one place only, the place where it is generated. Data may move instantly to wherever they are needed, in their most up-to-date form. When needed, they may be presented in their most useful format.

Understanding of manufacturing as a science, and the ability to bring to bear upon manufacturing information handling the power of modern computers, will enhance productivity and assure the attainment of corporate objectives.

# Chapter 15

# The Manufacturing Function—Reassembled

Having dissected the manufacturing function and examined it in great detail, it is now timely to reassemble the many bits that resulted from our analysis. We can better appreciate the whole, now that we are familiar with its parts. We shall see that manufacturing is indeed an unbroken continuum, from the initial perception of the need for a product straight through to the support of that product in the field. We shall see the complex connectivity of the many component functions, their interdependence, and their interaction. This emphasizes the wisdom of addressing the subject of manufacturing as a single entity, as we have done.

Although we initially approached the subject from the point of view of an enterprise principally engaged in manufacturing, we have concentrated on the manufacturing function per se, and not on the companion function of marketing. This chapter will, therefore, begin the reassembly with the manufacturing function as the point of beginning. Back in Chapter 2 there was a node tree diagram (Figure 2.2) that showed this point as Node A-12, "Manufacture the Product," one of the four major nodes under A-1, "Conduct a Manufacturing Enterprise." This point of beginning becomes the A-0 node discussed in Chapter 3. It was expanded into a full diagram (Figure 3.2), entitled "A0, Manufacture Products."

We found four principal parts of this manufacturing function: Develop the Products, Produce the Products, and Support the Service of Products in the Field, plus the overall managerial function of Manage Manufacturing. We examined each of these four functions and its subfunctions:

- Management of Manufacturing controls all the others. It has three subfunctions: Plan Projects, Make Project Schedules and Budgets, and Administer Projects. Figure 4.1 in Chapter 4 shows these activities.
- Develop Products originates, perfects, and designs the products, and maintains the designs throughout the life of the product. Figure 5.1 in Chapter 5 shows these activities. There will obviously be a managerial organization for Development, but it is not shown as a separate function.
- Produce Products converts procured raw materials into the finished products in accordance with the designs received from Development, and delivers the finished products to Marketing. Figure 6.5 diagrams this set of activities. There are six subfunctions, each of which is examined in a separate chapter of its own. Again, there is no separate explicit managerial function box in the diagrams, although there is implicitly a managerial function in the organization.
- Support the Service of Products in the field provides spare and repair parts to Marketing, and operating and instruction manuals. It is discussed, although not diagrammed, in Chapter 13.

The six subfunctions of Production are each discussed and examined in separate chapters:

Planning for Product Manufacture, Chapter 7;
Budgeting and Scheduling, Chapter 8;
Planning for Part Production, Chapter 9;
Provision of Resources, Chapter 10;
Obtaining Production Materials, Chapter 11; and
Conversion of Materials into Products, Chapter 12.

The four functions comprising Manufacturing, and their several subfunctions are shown in a node tree format in Figure 15.1, substantially as described above. A further step in the expansion of the diagram is shown for node A 36, Convert Materials into Products. These six subsubfunctions are all discussed in Chapter 12.

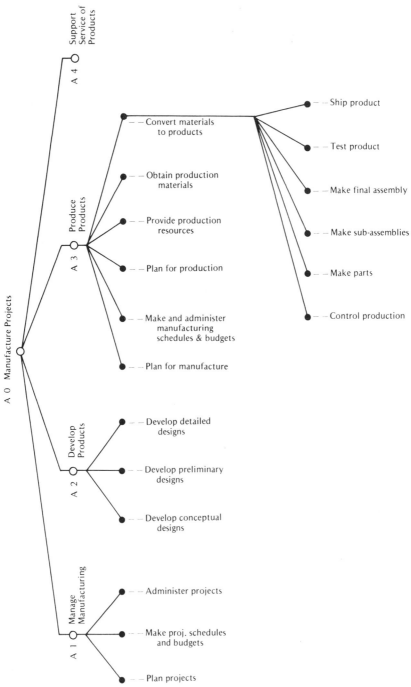

**Figure 15.1.** Node tree for manufacture products.

A 0  Manufacture Projects

Support Service of Products — A 4

Produce Products — A 3
- Convert materials to products
  - Ship product
  - Test product
  - Make final assembly
  - Make sub-assemblies
  - Make parts
  - Control production
- Obtain production materials
- Provide production resources
- Plan for production
- Make and administer manufacturing schedules & budgets
- Plan for manufacture

Develop Products — A 2
- Develop detailed designs
- Develop preliminary designs
- Develop conceptual designs

Manage Manufacturing — A 1
- Administer projects
- Make proj. schedules and budgets
- Plan projects

*178*

It is possible to carry the decomposition of the functional diagrams to much greater depths. Indeed, the Air Force's ICAM program has done so, and their work is most helpful. We too could go on expanding our functional diagrams to more and more levels of detail, but it is not within the scope of this book to do so. Particularly at the lower levels the structure is more and more dependent upon the product being manufactured, and thus would lose its generalized applicability. However, the reader may do so with considerable profit and contribution to his understanding of the science of manufacturing.

One of the rules of the IDEF modeling methodology, described in the Appendix and used in this book, limits the number of boxes in a diagram to no more than six. This is based on a sound fact that the human mind can grasp readily the relationships between six or fewer things, but tends to stumble if more than six are presented in one pictorial. Nevertheless, there are occasions such as the present when it is justifiable to break that rule and consolidate a number of boxes into one giant diagram, such as Figure 15.2 -- see cover foldout.

Figure 15.2 is not just a concatenation of Figures 4.1, 5.1, and 6.5 in the framework of Figure 3.2. There is some simplification, because arrows go direct from source to receiver without getting caught in the rules for labeling at the boundaries of the smaller diagrams. It is just this directness that makes Figure 15.2 so valuable. Please note two facts.

First, there is an extensive network of forward and reverse communication of both information and material. The flows may link adjacent boxes or boxes at the extreme ends of the chart. Each function has inputs and outputs, some times several of each. Arrows may split and feed the same data to several boxes, or merge to converge data on one box. Manufacturing is indeed a continuum, an unbroken fabric composed of many separate, but highly interdependent, functions.

Second, there is no obvious seam or division line in this network between development and production. For many years there has been a belief that these two functions were separable and only distantly connected. In the early part of this century, it was common to separate the two organizations in many ways. They reported to different and sometimes antagonistic vice presidents. They were frequently located in different buildings, or even in different cities. There were social and cultural differences in the staffs. It was thought that too much feedback from the shop to design would stifle innovation, or that too much freedom in design would demand unavailable production technologies, or drop a cascade of design changes on an otherwise stable factory.

Fortunately, the error of this separation syndrome has been recognized, and there is a strong move to bring the two into close proximity and free communication. The progress of some technologies, particularly the electronic and the data processing technologies, has forced an intimate collaboration and rapid communication. The educational, social, and cultural backgrounds of the two staffs have approached complete equality.

Figure 15.2 demonstrates the validity of this new perception of the component functions in the totality of manufacturing. There are no seams or cleavage lines in the fabric depicted there.

Another lesson to be learned from Figure 15.2 is the breadth of the impact of any change in any function or data flow. It will ripple out to the edge of the organization and affect many if not all the other functions. This points up the sensitivity of the operation to random disturbances, and the role of managment in controlling the dynamics of the operation.

Finally, note that the perimeter of the diagram has the same inputs, controls, outputs, and mechanisms as did the A 0 diagram (Figure 3.2) of which it is the expansion.

## MANUFACTURING ORGANIZATIONS

The difference between the functional structure of the science of manufacturing and the organizational structure of the people who staff the manufacturing enterprise has been emphasized several times. The organizational structure shows the controls (but not the inputs and outputs) of the manpower of the enterprise (but not of the products or projects). It shows who reports to whom, the spans of control, and hence of individual authorities and responsibilities. It does not treat the control of data or of materials, nor the dynamics of their movement through the manufacturing process, except that it implies that certain persons will properly plan and perform those controls.

In large manufacturing enterprises it is not uncommon to find that the organization charts contain a dozen or more levels between the chief executive officer and the operator on a machine tool. Solid lines show the administrative control from top to bottom, like a gigantic tree structure. Further, it is not uncommon to find dotted lines criss-crossing the administrative chart to show functional jurisdictions that control operational performances. If so, the result is a structure looking more like a spider web than a tree. These complex charts are attempts, and not very successful attempts, to display the functional structure of the manufacturing activity on a chart ill suited to that purpose. These attempts do show, however, the need for a better method to display the essential functional nature of manufacturing. I hope that the viewpoint and methodology presented in this book will offer a better approach.

When analyzing a manufacturing enterprise, it may be useful to map its organizational structure onto the functional structure we have developed. It is immediately apparent that the organizational units do not match the functional units. There are several excellent reasons.

- Several organizations may perform the same function. For example, there may be several separate plants producing the same kind of a product; developmental units for different products may all be doing the same type of design function.
- Not all enterprises perform all the functions on our generalized functional charts. For example, a company may have no in-house part fabricating capability whatever; they may instead buy all their products' parts made by others and simply assemble them; or a company may do no assembly work whatever; they merely make and sell pieces, such as electronic resistors or steel washers.
- In small organizations, one unit may perform several of the functions on the diagrams. For example, one small group working under one leader may do planning for manufacture, planning for production, and tool design (Figure 9.1).
- One function may be performed by several quite separate groups. For example, ⌐ group may design tooling in a different place and under different supervision from the groups designing products.

These, and other reasons for differences may prevail for long periods of time. There are other differences that are time limited, and may be expected to change from year to year. Leadership of an organizational unit must be assigned to a specific person, and people have differing levels of technical and supervisory competence. It is just as undesirable to assign a supervisory role to someone overqualified for the job as it is to assign it to someone underqualified. Therefore it is not uncommon to tailor the size of organizational units to the competence level of the available managers. As managers grow in skill they may be given larger and wider tasks to manage, and that changes the organization chart. Unfortunately, the converse is not always true. It is rare, when a manager's skill wanes or becomes obsolete due to changes in technology, that the span of his supervision is reduced, which not only would change the organization chart but might publicly degrade the manager.

Finally, tradition may dictate the form of organization rather than reality. Geographical names or persons' names may be used to designate an organizational unit, and the analyst must interpret before he tries to compare the chart to the functional diagrams. Deference and respect for an old and revered employee or officer may freeze an organizational configuration quite without regard to the things the organization is currently doing. Again, interpretation is necessary before the analysis may proceed.

Certain traditional names appear regularly on organization charts but do not appear in our functional diagrams. "Research and Development Division" is an example. The research function appears in Figure 15.2 in Box A21, titled "Develop Conceptual Design." The development function appears in the same figure in Boxes A22 and A23, but the detail design function may also cover the design of tooling as well as of products, as mentioned above. Traditionally, "R & D" is applied only to the company's products, not to their facilities. An organization called "Advanced Manufacturing Engineering" may assume the research function in the company's manufacturing technology. Perhaps one excellent way to get closer liaison between designers and producers would be to consolidate the design of the machinery and tooling of production with the design of the products to be made on the machinery. Conceptual Design (Figure 5.1, Box 2) may work hand in glove with Planning for Manufacture (Figure 6.5, Box 1), while Detailed Design (Figure 5.1, Box 3) may collaborate with Planning for Production (Figure 6.5, Box 3).

Research is frequently emphasized in corporate structures because of the competitive importance of research, as well as the prestige attached to that function by financial analysts and investors. Industrial research, as differentiated from academic research, made its appearance just before World War I, when General Motors Corporation, among others, set the precedent. Severance of the access to German research knowledge by that war showed how important it was to conduct applied research in industry, a lesson United States industry has never forgotten. The placement of the research function in our diagrams does not denigrate the importance of the function. It merely shows it in its sequential place in the total function of developing products. Note also that a company may conduct research in fields not directly connected to the development of some specific product. Those efforts are under the aegis of the Research Division, but do not appear in our functional structure, except that the fortuitous perceptions of new product concepts found during a more specifically directed effort may contribute to them.

Manufacturing Engineering is another name that appears on organization charts. When top management accepts from the R & D organization a product design, it is handed over to the Production Division, and more specifically to an organization commonly called Manufacturing Engineering, by whom it will be processed and then handed on to the "Factory." There is no one standardized format or job description for Manufacturing Engineering; every company has its own setup, and its own boundaries for this organization. It usually comprises those functions included in "Plan for Manufacture" and in "Plan for Production," shown in Figure 15.2, Boxes A31 and A33, but may also have other duties which we have included under "Provide Production Resources," Figure 15.2, Box A34.

It may be reasoned that if this group is going to specify how a part or product is to be manufactured, it should also specify the machinery, tooling, skills, and technology required. We have included these specification tasks in the two planning units, but the actual provision of the resources in a separate unit, as discussed in Chapter 10. For organizational purposes industry may place the selection, justification and purchase of new machinery in Manufacturing Engineering, working closely, the design of tooling may be part of Manufacturing Engineering, as may also be its in-house production, but its management and care will fall to others.

Manufacturing Engineering applies available technology to the manufacture of products already designed. It also may seek out and acquire new technologies and facilities to improve the manufacture of today's products. Further, it may do manufacturing engineering research to develop new methods which in turn may influence the detail design of future products. This is another form of research undertaken with the conviction that if there is a better way, it better be sought out, even though there is no predictable specific benefit known. It is research directed at the company's internal activity in producing products, and parallels the research discussed above directed at the functions performed by the company's product when in use in the field.

Purchasing is a special art, usually accorded a separate organizational spot. We have placed it in "Obtain Production Materials," Figure 15.2, Box A35. Allied with it, but frequently assigned to some other part of the factory organization, are the Receiving Department, Incoming Inspection, and Raw Stock Storage.

There will ordinarily be a managerial organization in charge of the entire area we have designated as "Produce Product." Included in it will be the function we called "Make and Administer Schedules and Budgets," Figure 15.2, Box A32, and possibly some of the functions we have called "Control Production," Figure 12.1, Box 1. Similarly there will probably be a managerial organization in charge of the Development function.

Assembly and Shipping may be a single organization when very large products are involved. In other cases Shipping and Receiving may be lumped into one organization, probably because they both must be located near the truck dock, and both deal with outsiders.

There will also be other units treated as separate but which are functionally part of he master diagram. Materials Handling is an example. Every time a piece of material has to be moved from one site to another, the services of a vehicle (a truck, chute, conveyor, or even a human with a tote pan) and usually an operator are required. They serve all stages of the production function, but usually are organized as an "overhead" department, meaning that the cost of moving some specific piece of material is not allocated by Accounting to the cost of producing that part, but is spread over all the output, along with the cost of all materials movements. Equipment maintenance, security, and other such functions may also be treated as "overhead".

Quality Assurance and Control are two other functions that usually get separate organizational status, even though the maintenance of quality is as basic a function as is detailing or packing; see Chapter 5 for a discussion of this. Inspection operations are just as much a part of the operations sheet as turning or drilling, as we saw in Figure 9.3, but because one man may inspect and reject the work of another, it is common practice to put the inspectors in a separate organizational unit reporting to a level equal to or above the level of the head of the producing unit. This creates a separate box on the organization chart, even though the inspection or test is properly a production function. The same applies to testing of the finally assembled products.

On the other hand, quality assurance is a function of planning the manufacturing process so that a greater level of acceptable product will be generated in the first place, rather than trying to cull out and correct mistakes. Quality Assurance work is performed primarily in the Preliminary and Detailed Design units in the Development function, and by the two planning functions of the Production unit, even though it may be organizationally allied with Quality Control.

Analysis of an organization by comparison to the basic functional structure of the science of manufacture will help to understand what is going on. It will also permit the analyst to transfer and use prior knowledge gained in other manufacturing enterprises to an organization under study, using the basic functional structure as a common point of reference.

## SUMMARY

Having dissected the manufacturing function and examined it in great detail, the parts may be reassembled into the whole from which they were derived. We can better appreciate the whole, now that we are familiar with its component parts.

A node diagram and a function chart of the entire manufacturing activity is presented, to consolidate the total concept. It may be seen that manufacturing is indeed an unbroken continuum, from the initial perception of the need for a product straight through to the support of that product when it has been developed, produced, and placed in use.

Manufacturing is a very complex activity, the many parts of which are interconnected and interact through a network of data and material flows between the functions. There is no obvious, or even hidden, line of demarcation between the development activities and the production activities, as was once thought.

Owing to the complex interaction between the component functions, the whole is sensitive to accidental perturbations in any one part.

Analysis of any given real organization will be facilitated by comparison of the organization to the functional structure of manufacturing. There will be differences for many reasons, some valid permanent differences and some dependent only upon the variations from person to person, from time to time.

Terminology of the organization will therefore differ from the terminology of the functional science we have developed, but should cause no difficulty in its useful application.

# Chapter 16

# Conclusion

Manufacturing is a science. Of course, it has always been a science, even though for millennia it was regarded and practiced as an art. However today we *practice* it as a science, and enjoy the benefits of this deeper insight and control. The key to increasing the productivity of our manufacturing enterprises is this understanding of the process as a science.

In this book we have developed the structure of the manufacturing process in some detail, using the techniques of structural analysis, but not to the exhaustive detail which is possible when necessary. The structure has been diagrammed in such a manner that it is easy to comprehend. We have identified the component parts, defined them, and traced the relationships between them. The structure has been defined in a mathematically rigorous manner; this precision of definition will permit us to utilize the analytical techniques well known to science.

When we compare this generic structure of manufacturing to a specific real enterprise, differences will appear. Some may be quite proper differences, as for example, when operating a job shop producing items to other peoples' designs, in which case there may be no product development unit in the organization. The remainder of the generic structure will still be directly applicable to the case in hand.

Other discrepancies in the structure will raise appropriate questions for further investigation.

When we are dealing with a specific enterprise we can quantify the various functions – the times, the speeds, the volumes, the data formats, and the like – and then apply scientific analytical techniques to determine the controlling parameters, and their impacts on each of the components and their communications. Once we have done this, we can by simulation predict the performance of the enterprise under a variety of assumed conditions, and optimize our course of action.

I believe that the structure of manufacturing set forth in this book is indeed the common infrastructure of all manufacturing enterprises, regardless of the product being manufactured. One has only to look below the superficial differences and the trade jargon to find this generic structure. Furthermore, I believe that it is reasonably invariant with respect to time, so that we may rely upon it for guidance in the future. This is not to say that, with the passage of time and the accumulation of large amounts of experience, the structure may not evolve. For the time being, we may use this structure to help in transferring the lessons of the past for use on the problems of the present.

We have defined manufacturing as the conversion of naturally occurring material into desired end-products. We have seen the tremendous complexity and the almost universal interaction of the many component functions of manufacturing. It follows from this that if we change any one element of this structure, we will in some way affect every other element. It further follows that we cannot treat any one portion of the structure without due concern for the remainder. The truth of this interdependence has been painfully demonstrated whenever it was ignored. Therefore, we must remember that the scope of manufacturing extends from the perception of the need for an end-product, right through the life cycle of that product, until the last unit goes out of use in the field.

Nevertheless, to keep the size of this book within bounds we have discussed only the "discrete parts industries," and only those parts of those industries which deal with making the product as distinct from marketing the product. To protect ourselves from charges of tunnel vision, we have noted the points of interaction, and the mutual influences at those points, with marketing and enterprise management, and with the extractive and process industries. Within our selected area of interest, we have treated manufacturing as an indivisible continuum.

All of the functions, and all of the communications or the flow of materials between them, may be expressed as data. At each point, data are being generated, or transformed, or transmitted. Included in "transmitted" or both transmissions over space, as well as storage over time. This universal presence of data is a common thread that knits together all the functions and communications, all the parts and all the persons, involved manufacturing.

Furthermore, these data are suitable for digital electronic processing. Fortunately, we are today in possession of powerful and adaptable tools for data processing, and the speed and versatility of these computers is increasing at a pace that seems to increase exponentially. Computing power is thus another powerful force for the reintegration of all the diverse bits of manufacturing as we knew it just a few years ago. Data processing technology requires a mathematically rigorous definition of the terms in which data is couched, and the codes, grammar, and syntax in which it is expressed. This is another driving force in the adoption of a carefully conceived and written structure of manufacturing.

We can conclude that manufacturing is a science whose structure we understand. It can be analyzed, and having analyzed it we can make predictions. We can determine what the basic parameters are, and how they should be measured. When we can measure, we can control manufacturing. When we can control, we should be better able to succeed.

# Appendix

# IDEF Graphical Modeling Technology

The graphical modeling methodology used in this book has been adapted from the IDEF$_0$ methodology developed by the Air Force Integrated Computer Aided Manufacturing Project for a similar purpose. The history of the ICAM program and of one of its best products, the IDEF system, is of interest.

The Materials Laboratory of the Air Force Systems Command is loacted at Wright-Patterson Air Force Base, Dayton, Ohio. It was charged in 1977 with the task of carrying out the Integrated Computer Aided Manufacturing (ICAM) program. The purpose of the program is to expedite the development of advanced manufacturing technology in order to reduce the cost and increase the speed of delivery of aircraft for the Armed Services. This ambitious goal needed a common "baseline" communication system around which to plan, develop, and implement the system improvements. A model of manufacturing, and a language for its communication, were needed.

The specifications called for the expression of manufacturing operations in a natural straightforward way, for conciseness, and for rigor and precision. This was not achieved overnight! An initial effort had been taken in 1974 in a predecessor program called Air Force Computer Aided Manufacturing (AFCAM). The subcontractor on that project was

SofTech, Inc., Waltham, Massachusetts, a software house that had developed a methodology called Structural Analysis and Design Technique for just this type of purpose. Boeing Commercial Airplane Company as prime contractor furnished the technical background, and a model was built. The model is called "IDEF," standing for ICAM Definition system. It was referred to as the "architecture of manufacturing." The whole concept was alien to the aerospace industry's usage, and was not joyously received!

The ICAM program was started at the direction of the Deputy Secretary of Defense, W. P. Clements, Jr., April 11, 1975. The prime responsibility for the architectural modeling was this time given to SofTech, who enlisted seven large aerospace manufacturers, including Boeing, as subcontractors. The model generated by this group of people was perfected out of a consensus of the group, thus showing that there is a common fundamental structure in manufacture, even though its embodiment in the several firms varied considerably. My role in the project was to test the model, as it was being developed, against practice in nonaerospace firms. I showed that indeed the fundamentals of manufacturing were the same in aerospace as in any other manufacturing industry, granting of course that nomenclature was properly adjusted to the various trade usages and jargons.

The hierarchical structure of IDEF was extended downward in detail to many levels, and upward to orient the entire activity within the interests of the Air Force, and in turn within the concerns of national defense. The main structure of the IDEF model is considered complete (1982), but as applications take place, modifications and extensions are constantly being made. One version, the $IDEF_0$ model deals with the functional structure of manufacture. Another version, $IDEF_1$, deals with the nature of the data handled by the systems, a subject of great importance in computerized versions. The third version, $IDEF_2$, deals with the dynamics of manufacturing systems, the flow of materials and information in a time frame, and subject to the constraints of availability of men, materials, and machines. Handling the data necessary to build and use such a model is a monstrous task; to assist, it has been computerized itself.

Textbooks and manuals on IDEF may be obtained, subject to certain limitations, from the Materials Laboratory, Air Force Wright Aeronautical Laboratories, Air Force Systems Command, Wright–Patterson Air Force Base, Ohio 45433.

This book does not require the depth of detail of the work done for ICAM, but it will use the principles as we develop our own model and its illustrative diagrams. The model of manufacturing used in this book differs from the ICAM model only in that it is written from the point of view of a company engaged in manufacturing some product, rather than the point of view of the Air Force engaged in the supply and use of military aircarft. The most significant differences are in the managerial levels of the model.

The model is, I believe, applicable to the manufacture of any product. It has been tested in a wide variety of industrial environments, and proven workable. It has been adopted voluntarily and used productively, which speaks strongly to the inherent rightness of the method. The method has even been used to structure and manage projects intended for the generation of new data processing techniques.

Not surprisingly, there is a strong parallel between the creation of models by this method and the creation of written materials, such as this book. In good writing, an author first sets forth his thesis – a short statement containing the essence of what he intends to present. This is then expanded into half a dozen or more sentences which collectively restate the thesis, but in much more detail. Each of these sentences is again expanded into further detailed exposition of the thought. Ultimately the author arrives at a collection of sentences each of which will become a paragraph in the final essay. That paragraph then sets forth everything necessary to support its thesis sentence, but nothing else. The result of such structuring is a tightly written, persuasive document using language efficiently and effectively.

In a similar manner, the IDEF modeling procedure sets forth, in one diagram, the scope of what is to be modeled, and the major functions involved in that activity. Then each of these is expanded into a diagram of its own, showing the functional structure therein. In turn, each of these functions is expanded, and so on, until the desired amount of analysis and exposition has been achieved. The resulting hierarchy of diagrams is a lucid, easily comprehended diagramatic model of the original thesis statement.

The basic concepts of the IDEF graphical modeling techniques can easily be understood. First of all, the model consists of diagrams, text, and a glossary. The diagrams are the two-dimensional models discussed in Chapter 1. Each one is simple enough to fit on a single page of a book. There is a labeling system that acts as an index and a guide to the relations between the collection of diagrams. The text is a discussion of the elements of function or data transfer shown on the diagram. It is closely keyed to the diagram to ensure that there shall be no ambiguity. The glossary acts as a final definitive authority for what each word means as used in the diagrams or text, within the specific context of the model. This is most important, because industrial use of the English language and its words is, to say the very best, imprecise. The model builder, therefore, nails down his precise meaning, and the reader must respect that definition while using the model.

A manufacturing function is represented by a box, and the interfaces between the functions are represented by arrows. Each box may have an input, a control, an output, and a mechanism interface, represented by arrows entering or leaving the box. Figure A.1 shows this arrangement. Input arrows always enter on the left-hand side, control arrows from the top, mechanism arrows from the bottom, and the output arrows always leave from the right-hand side.

The function in the box is constrained to operate only on the input received, only at the direction of the control, and only by means of the mechanism indicated. The output of a box is constrained to go to another function, or functions. Figure A.2 illustrates this with a specific example of a machine shop function – machining a shaft. The input is the raw material for the shaft, in this case a piece of bar stock. The control is the detailed part drawing showing exactly what is expected to be produced. The mechanism is a lathe. The output is a finished shaft. The input is converted into the output, and goes with it. However note that neither the mechanism nor the control is consumed by the function; they both stay with the box.

A diagram contains a number of boxes, connected by arrows. Figure A.3 shows a diagram containing four boxes. (It is drawn from one of the diagrams we will be developing in the model, and is introduced here only to help expound this paragraph.) It shows two inputs, one control, and one output. There are no external mechanisms shown, purely for the sake of simplicity. Note that some of the output of Box 1 becomes input to Box 2, while another output of Box 1 becomes a control for the remaining boxes. Note also that some of the outputs from both Boxes 1 and 2 feed back to become controls for Box 1. The output of Box 4

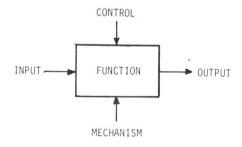

**Figure A.1.** Modeling concepts.

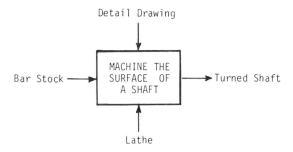

**Figure A.2.** Example of a function box.

supplies the mechanisms needed for Boxes 1, 2, and 3. Thus we can see how a single diagram can portray a whole complex set of relationships between its component functions, and the relationship of the group of functions to the external world.

The most important part of the IDEF system concept is the gradual exposition of detail by expansion of the contents of the boxes on a diagram into lower level diagrams. Each box becomes the "parent" of a diagram on the next lower level. Figure A.4 illustrates this in a perspective diagrammatic form. A single box in the highest level diagram — a diagram numbered A-0 (read this as "A minus zero") — is expanded on the next level below into a diagram with several boxes and arrows. This diagram is numbered A0 ("A zero"). Each of these boxes in turn become the parent of another diagram on the next lower level. In this figure, only the fourth of the four boxes is shown expanded into diagram A4. In turn, the second box in that diagram is shown expanded into a diagram numbered A42 on a still lower level. The "A" in this nomenclature designates an "activity."

This expansion may be thought of as the trunk of a tree branching out into limbs and in turn into smaller branches. This process of gradual exposition has become known as the process of "decomposition." The higher level diagrams have a great deal of implied information, while the lower level diagrams have less information but display it in greater detail. The process may be carried as far as necessary to explore any facet of the total manufacturing activity. However no matter how far it is carried, each diagram contains no more information than the mind can grasp at one glance. The rule is that there shall be no more than six boxes in a diagram.

A box represents a function that is described by a verb or a verb phrase that is written in the box. The boxes are numbered by adding a digit to the diagram number; the box number becomes the diagram number when it is expanded into a diagram. The number of digits in

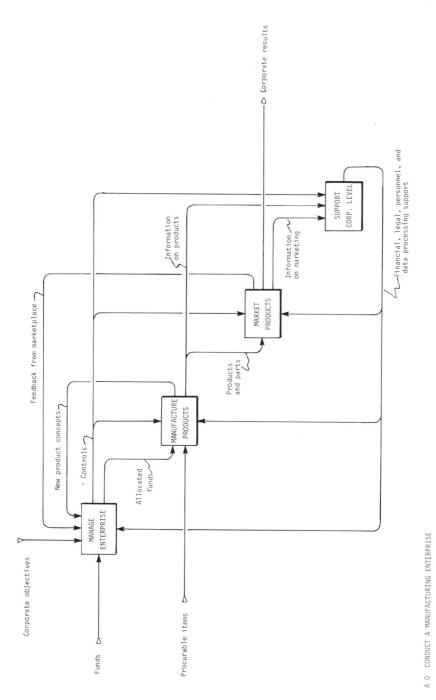

A 0 CONDUCT A MANUFACTURING ENTERPRISE

**Figure A.3.** An IDEF diagram, showing how the boxes, representing functions, and the arrows, representing flows of data or materials, are brought together to form an IDEF diagram. This diagram is not a part of the model of manufacturing to be developed in the text; it is merely introduced here to illustrate the principle.

194

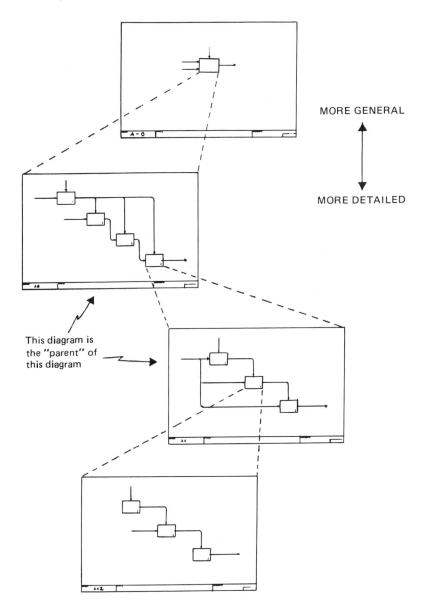

**Figure A.4.** Decomposition of diagrams.

a box number is indicative of the level of the diagram on which it appears. The functions may represent any act in manufacturing, from the concrete – such as cutting metal – to the abstract – such as classifying.

The arrows entering a box are numbered and coded by their points of entry or departure. Thus I1, I2, . . . are the various input arrows; C1, C2, . . . are the control arrows; O1, O2, . . . are the output arrows; and M1, M2, . . . are the mechanism arrows. A box number and one of these "I C O M" numbers will identify one end of any arrow. When a box is expanded into a diagram, that diagram must have the same inputs, outputs, controls, and mechanisms as does its parent box – no more, no less. Arrows entering a diagram are identified in its margin by the ICOM numbers of the parent box. This rule is rigorously applied, which contributes to the integrity of the model.

Arrows that connect boxes represent data (objects or information) needed by or produced by the functions. Arrows are described by a noun or a noun phrase written beside the arrow or connected to it by a "squiggle." The term "data" is used here to designate anything which may be used to transfer information. For example, a template used to guide a cutting tool embodies the information which defines the geometric path to be followed by the tool; a drawing of a part defines the exact configuration which the part is to have; a route sheet specifies a sequence of events. All these are forms of data to be transferred.

Input arrows show the data needed to perform the function; it will be consumed by the function and converted into the output. They show WHAT will be USED. Output arrows show the data which will be created when the function is performed; they show WHAT was PRODUCED. Control arrows describe the conditions or circumstances that govern the function; they are not consumed by the function. Every box has at least one control arrow, and some have many controls. Mechanism arrows show what person or device carries out the function. They show HOW it is performed.

Arrows do not show the sequence of flows. When there is a feedback, or iteration, or looping, or continuous processes, or overlap in time, the diagram will show that. Arrows may branch or join, in which case it is necessary to be very careful with arrow labeling.

Node numbers are assigned to each diagram, and to boxes within each diagram. The node number for a box becomes the node number for the diagram into which it may be decomposed. Thus a hierarchy of diagrams is formed. It may be presented as a node tree, or as an indented listing similar to a bill of materials. Either acts as a convenient index to the compiled diagrams, but it is easy to see from the sample node tree shown in Figure A.5 how rapidly the tree diagram would get un-

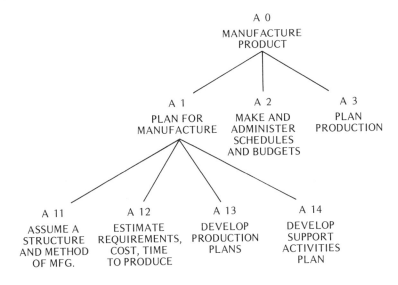

Diagrams form a "Hierarchy" shown by a Node Tree

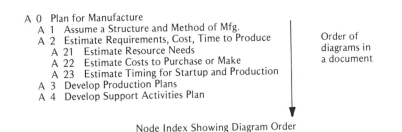

Node Index Showing Diagram Order

**Figure A.5.** Node tree and a node index (for illustration only).

manageably sprawled out when more than two or three levels of decomposition are presented.

It is important to repeat again: These diagrams are NOT organization charts. A function may be performed by one person or many; they need not all report to the same supervisor. One person may perform several functions, particularly in a small company. They need not fall in adjacent boxes in the diagrams. What these diagrams DO represent are the discrete functions inherent in the science of manufacture.

# Glossary

Precise definitions of the terms used to describe the function boxes and arrows in the diagrams in this book. References in parentheses are to the locations in the diagrams.

**Accepted Products:** Products which have passed the final test and are forwarded to Shipping (A36).

**Administer Projects:** Monitor and direct the execution of a project (A36).

**Assembly:** A unit of the product consisting of separate parts. See **Final Assembly** and **Subassembly** (A36).

**Authorization:** A mandate from management to perform subordinate functions, as Authorization to manufacture, Authorization to market, Authorization to support, etc. (A-1, A-0, A0, A1).

**Bill of Materials:** Hierarchical listing of all parts comprising a product, assembly, or subassembly. See **Design Bill of Materials, Final Manufacturing Bill of Materials.**

**Change Orders:** Notices of changes made in product designs in response to requests from manufacturing, or to modify or enhance the product's characteristics and performance (A0, A2, A3).

**Conceptual Design:** The design of a technically practicable product configuration (A2).

**Conduct a Manufacturing Enterprise:** The primary functional concept around which this whole study is based (A-2, A-1).

**Control Production:** Management of the fabrication, assembly, testing, and shipment of products (A36).

**Controls:** Distributions of authority and responsibility to subsidiary functions (A-1).

**Corporate Objective:** The purpose of the enterprise; controls the enterprise's existence and activity (A-2, A-1).

**Cost and Status Reports:** Reports to higher levels in the organization of accumulated cost to date and state of progress of work (A3, A36).

**Design Bill of Materials:** The version of the Bill of Materials that defines the product as it emerges from Development (A3).

**Design Change Requests:** Changes in the design of a product (usually in the detailed design of parts) requested by Production, Support Services, or Marketing to facilitate fabrication or assembly, conserve materials or time, increase reliability and decrease service, etc. (A0, A2, A3).

**Design Concept:** Output of the conceptual design phase (A2).

**Design Information:** Access to the product design, for the guidance of persons outside the development department (A0).

**Detailed Design:** Output of the final design stage. Complete definition of all the component parts, and their interrelationships in the ultimate product; layouts, details, bills of materials (A2, A3).

**Detailed Design Data:** Information concerning the detailed design, fed back from Detailed Design to Preliminary Design, for their information and guidance (A2).

**Develop Conceptual Design:** The initial step in development, directed to using or acquiring the basic concepts of the product design; research, experimentation, modeling, etc. (A2).

**Develop Preliminary Design:** The conversion of design concepts into initial product configurations; experimental testing in simulated use (A2).

**Develop Detailed Design:** The final step in development: preparation of layouts, details, bills of material; final testing. Also maintenance of the design by effecting redesign as necessary (A2).

**Develop Products:** Create and maintain product designs (A0).

**Development:** The creation and maintenance of product designs. Contrast with **Production** (A0).

**Directives:** Instructions from Manufacturing Management to perform subordinate functions, e.g., develop, produce, support (A0, A1, A2, A3).

**Enterprise:** An individual, a partnership, a company, a corporation, or a government agency engaged in manufacturing (A-2, A-1).

**Equipment:** Machinery and subsidiary devices; may be general or special purpose machines (A-1, A-0, A0, A3, A36).

**Experience:** Information or know-how, transferred from the past to the present by those concerned (A-1, A-0).

**Facilities:** Office, factory, and warehouse buildings, and similar assets (A-1, A-0, A0, A3).

**Feedback:** Intelligence from one function to another function earlier in the manufacturing sequence, pertinent to their function (A-1).

**Final Assembly:** A group of parts and subassemblies, fastened together to make the deliverable end-product. Contrast with **Subassembly** (A36).

**Final Manufacturing Bill of Materials:** The bill of materials that governs the production process (A3).

**Funds:** Appropriated monies, a portion of resources (A-1, A-0, A0, A1).

**Information on Marketing; Information on Products:** Information for the Corporate support function regarding the activities and output of the Manufacturing and Marketing functions (A-1).

**Inspection:** The examination of a part to determine the degree of its conformance to the design. Accepts or rejects the workmanship of material, but does not alter its configuration. See **Test.**

**Instructions:** Written material of all sorts describing installation, operation, repair, and maintenance of products. Also: detailed instructions for fabrication and assembly of parts (A0, A3, A36).

**Make Assemblies, Make Subassemblies:** Aggregation of parts into assemblies. See **Assembly** (A36).

**Make Parts:** The physical conversion of raw materials into parts which will be used in the product, or as spare parts. Involves fabrication and inspection (A36).

**Make Project Schedules and Budgets:** See **Project Master Schedule.**

**Make and Administer Manufacturing Schedules and Budgets:** See **Production Schedules and Budgets** (A3).

**Manage Enterpirse:** The managerial control for the entire enterprise, including manufacturing and marketing (A-1).

**Manage Manufacturing:** The management of the manufacture of products as distinguished from the marketing of the products (A0, A1).

**Manufacture Products:** That portion of the enterprise's efforts directed to the conversion of raw materials into saleable products (A-1, A-0, A 0).

**Manufacturing:** The conversion of raw materials into desired products.

**Manufacturing Capabilities:** Statements of Production's abilities and limitations in fabrication, assembly, inspection, and test. Does not include the current availabilities of facilities or equipment (A0, A2, A3).

**Manufacturing Information:** Detailed information concerning manufacturing methods, dimensions, cost, and time involved, etc., useful to design and support functions (A3, A36).

**Market Products:** That portion of the enterprise's efforts directed to the selling of its products to users (A-1).

**Marketplace:** Those portions of the world which are potential buyers of the enterprise's products.

**Marketing:** The transfer of manufactured products to the marketplace, and the support of those products in the hands of users. Includes the perception of needs for products, the establishment of marketing policies, advertising, sales, and services (A-1).

**Marketing Objectives:** Policies and programs for the governance of the activity of the marketing function (A-1).

**Master Schedule and Budget:** The controlling time and cost expenditures planned for the salient milestones in the manufacturing process. See also **Production Master Schedule, Project Master Schedule.**

**Materials:** Procured items including both raw materials and components, to be worked in to the finished product (A3, A36).

**Materials Specifications:** Information for the Obtain Materials function concerning the materials requirements for work orders being entered into production. (A3).

**Mechanism:** The person or device that carries out the function named in a box of the diagrams (A-2).

**New Product Concepts:** A concept for a new product, arising during development but outside the assigned product definitions and specifications. A perceived opportunity for a desirable product (A-1, A-0, A0, A2).

**Obtain Production Materials:** See **Materials** (A3).

**Parts:** Individual pieces, the result of a manufacturing process prior to any assembly. Parts may be assembled into products, or sold for spare and repair purposes. Also, parts may be the end result of manufacturing, as in the case of coins, door keys, paper clips, etc. (A-1, A-0, A0, A3, A36).

**Personnel:** The work force of all ranks, operative and managerial, direct and overhead (A-1, A-0, A0, A3).

**Plan:** The formalized procedure for manufacturing; to create a plan (A1, A3).

**Plan Change Requests:** Requests for changes in the planning for a project or product, referred to the management of manufacturing to overcome problems not otherwise surmountable (A0, A1, A2).

**Plan for Manufacture:** The overall planning for the manufacture of a product. Contrast with **Plan for Production** (A3).

**Plan for Production:** The planning for the fabrication of individual parts and their assembly (A3).

**Plan Projects:** The creation of a plan of procedure for the manufacture of a specific project (A1).

**Policies:** Managerial decisions governing procedures and strategies (A-1).

**Preliminary Design:** A functionally practicable configuration of a product design, proven, but not yet engineered for production, cost reduction, etc. (A2, A3).

**Preliminary Design Data:** Information concerning the preliminary design, fed back from Preliminary Design to Conceptual Design for their information (A2).

**Problems:** At all levels, perceived deviations from plans, budgets scheddules, fabrication or assembly, either actual or imminent (A1, A3, A36).

**Procurable Items:** Input materials available to the enterprise by the use of funds (A-1, A-0, A0, A3).

**Procured Items:** Materials and components purchased by the enterprise for incorporation either in the product or in the production of tools, tooling, facilities, etc. (A3, A36).

**Produce Products:** Convert procured materials into finished products, in accordance with the product designs; production (A0, A3).

**Products:** Products are the results of the manufacturing process. In the discrete parts industries, products are assemblies of parts. For example, automobiles, typewriters, shoes, and books (A-1, A-0, A0, A3, A36).

**Product Definitions and Requirements:** Detailed statements of a product's required characteristics, performance parameters, cost, etc. (A-1, A-0, A0, A1, A2).

**Product Designs:** Layouts, details, bills of materials and other documentation needed to completely define the product (A0).

**Production:** The physical conversion of raw materials into products. Contrast with **Development** (A3, A36).

**Production Instructions:** See **Instructions** (A3, A36).

**Production Master Schedule:** The controlling times planned for the major steps in the production function (A3).

**Production Resources:** The necessary equipment, tooling, tools, personnel, and technology necessary for the production of work orders entered into production (A3).

**Production Schedules and Budgets:** The controlling time and cost expenditures for each detailed step of the production process (A3, A36).

**Programs:** Managerial decisions governing specific activity units of specific projects or products (A-1).

**Project:** A unit of manufacturing activity, identified separately for accounting or managerial purposes. May be a research project, a developmental project, or a production project (A1).

**Project Master Schedule and Budget:** The grand plan for the accomplishment of a project, setting forth times and costs at the major milestones in the plan. See **Master Schedule and Budget, Production Schedule and Budget** (A1).

**Prototypes and Models:** The various experimental devices used in research and development laboratories to perfect product designs. Also early versions of new products, for testing functional performance, reliability, etc. Only the knowledge gained from these items appears in the final output of the enterprise (A0, A2, A3, A36).

**Provide Resources:** See **Resources** (A3).

**Purchased Parts and Components:** Items to be incorporated in the product that are not made in-house (A36).

**Reconfiguration Request:** Request for a change in the conceptual or the preliminary design to optimize the following design stages (A2).

**Reports:** See **Status Reports** (A-0, A0, A3).

**Rejected Products:** Products which fail to pass the final test for acceptance and must be returned to assembly for disassembly and repair (A36).

**Requests:** At all levels, appeals for materials, equipment, resources, tooling design, design changes, etc., addressed to the normal source thereof (A0, A3).

**Requirements:** At all levels, specific elements of the plans, designs, etc., which become controls for subsequent functions, e.g., Production Planning Requirements, Resource Planning Requirements (A-1, A3, A36).

**Resources:** Money available for the operation of the enterprise, plus rights such as patents, contracts, or privileges. Also materials available through the use of money (A-2, A3, A36).

**Resources for Product Support:** Items used for services, repair, adjustment or rebuilding of products in the field (A-0, A0, A3).

**Resources for Production:** Includes facilities, equipment, tooling, personnel, technology (A3).

**Results:** The tangible accomplishments and yields from operation of the enterprise (A-2, A-1).

**Schedules and Budgets:** See **Master Schedule and Budget Production, Schedule and Budget, Project Schedule and Budget** (A36).

**Schedule Requirements:** The constraints for building the production schedule (and budgets), dictated by the plan for manufacturing (A3).

**Ship Products:** The packaging and delivery of completed products for transportation to the user (A36).

**Spare Parts:** Parts made for distribution to users of the products in the field, to maintain the products in operation. Sometimes: Spare and Repair Parts (A36).

**Specifications:** Detailed information concerning the requirements necessary to meet the plan for production; specifically, for resources and materials used in production (A3).

**Status Reports:** Reports to higher levels in the organization, giving costs to date and accomplishments vis-á-vis schedules; also any plan changes under consideration (A-1, A0, A1, A2, A3, A36).

**Subassembly:** A group of parts (purchased or made in-house) positioned and fastened to make a unit of the ultimate product. Contrast with **Final Assembly** (A36).

**Support (Enterprise Level):** Financial, legal, personnel, data processing, etc. services essential to the enterprise's operation, but which specifically do not use manufacturing or marketing skills (A-1).

**Support Requirements:** Policies and programs for the governance of the Corporate Support function (A-1).

**Support Service of Products:** The provision of instruction and maintenance manuals, spare and repair parts, and resources for support of the products in the field. Also the feedback of field information (A0).

**Technology:** The method of manufacturing (A-1, A-0).

**Test, Test Product:** Evaluation of an assembly to determine its conformance to designed charactertistics. Accepts or rejects an item, but does not alter its configuration. See **Inspection** (A36).

**Tool Design Request:** Request for the design of needed tools, tooling, or equipment to be designed and made in-house (A0, A2, A3).

**Tooling:** Cutters, jigs, fixtures, dies, templates, forms, and gauges essential to the action of the equipment and personnel in manufacturing operations. Also hand tools, benches, shelving, and the like (A-1, A-0, A0, A3).

**Tooling Specifications:** Information to the Resource Provision function concerning the tooling that will be necessary for work orders being entered in the production schedule (A3).

# Index